AF355998

Advances in Textile Materials Chemistry

Advances in Textile Materials Chemistry

Editors

Ana Sutlović
Sanja Ercegović Ražić
Marija Gorjanc

 Basel • Beijing • Wuhan • Barcelona • Belgrade • Novi Sad • Cluj • Manchester

Ana Sutlović
Faculty of Textile Technology
University of Zagreb
Zagreb
Croatia

Sanja Ercegović Ražić
Faculty of Textile Technology
University of Zagreb
Zagreb
Croatia

Marija Gorjanc
Faculty of Natural Sciences
and Engineering
University of Ljubljana
Ljubljana
Slovenia

Editorial Office
MDPI AG
Grosspeteranlage 5
4052 Basel, Switzerland

This is a reprint of the Special Issue, published open access by the journal *Molecules* (ISSN 1420-3049), freely accessible at: www.mdpi.com/journal/molecules/special_issues/Textile_Materials_Chemistry.

For citation purposes, cite each article independently as indicated on the article page online and using the guide below:

Lastname, A.A.; Lastname, B.B. Article Title. *Journal Name* **Year**, *Volume Number*, Page Range.

ISBN 978-3-7258-1666-8 (Hbk)
ISBN 978-3-7258-1665-1 (PDF)
https://doi.org/10.3390/books978-3-7258-1665-1

Contents

About the Editors

Ana Sutlović

Prof. Dr. Ana Sutlovic works as a professor at the University of Zagreb Faculty of Textile Technology. Her areas of expertise include textile dyeing and textile printing with natural and synthetic dyes. Over the last few years, the following research topics have been highlighted: natural dyes, digital ink-jet textile printing, 3D printing, thermochromic dyes, and waste water treatment. She is engaged in the teaching process and in the professional work that is carried out within activities of the TTF for the needs of the textile industry. So far, Ana Sutlović has co-authored 4 chapters in scientific books, 40 papers in scientific journals, and 121 conference scientific papers. During her career, she actively participated in 17 scientific research projects and professional projects.

Sanja Ercegović Ražić

Prof. Dr. Sanja Ercegovic Razic works as a professor at the University of Zagreb, Faculty of Textile Technology. After attaining her Master of Science degree, she received her PhD in 2010. She was the leader of 4 projects and collaborated in 11 scientific and professional projects. She has published 30 articles in journals and presented more than 60 papers at conferences. Her research focuses on developing functional and protective properties by modifying textile materials using environmentally friendly plasma technology.

Marija Gorjanc

Assoc. Prof. Dr. Marija Gorjanc works as a professor at the University of Ljubljana, Faculty of Natural Sciences and Engineering (FNSE), where she also received her PhD in 2011. She received the International TRIMO Award for her PhD thesis on the plasma modification of textiles for the application of silver nanoparticles. She was the leader of five bilateral projects, led activities in work packages of the EU project APPLAUSE and a Slovenian national research project centred around the ecologically friendly in situ synthesis of ZnO nanoparticles for the development of protective textiles. Her research deals with the development of environmentally friendly and sustainable colourful textiles with functional and protective properties.

Preface

The aim of *Advances in Textile Materials Chemistry* was to provide an overview of scientific research in the field of developing advanced textiles that are multifunctional or have targeted properties, emphasising the use of modern and environmentally friendly processing methods. The editors would like to thank all the authors who have contributed and published their valuable research results in this reprint. In keeping with its title, the reprint covers a wide range of topics. First, it refers to the research centred around the application of plasma in the paper *Mechanisms Involved in the Modification of Textiles by Non-Equilibrium Plasma Treatment*, and it presents the fundamentals of plasma penetration in textiles and illustrates mechanisms that lead to the appropriate surface treatment of fibres inside the textile. The textile industry is one of the major sources of pollution, with consumption and water pollution standing out. For this reason, great importance is given to environmentally friendly textile processing technologies, as can be seen in the important research paper *Influence of Cotton Cationization on Pigment Layer Characteristics in Digital Printing*. Considering the fact that multifunctional advanced properties are expected from today's textiles, the contribution of research through manuscripts *Silver Nanowires and Silanes in Hybrid Functionalization of Aramid Fabrics* and *Antibacterial Properties of Non-Modified Wool, Determined and Discussed in Relation to ISO 20645 2004 Standard* is important. In recent decades, there have been scientific research efforts aimed at reviving and commercialising the use of natural dyes in textile dyeing. This area is represented by the papers *Influence of Cotton Pre-Treatment on Dyeing with Onion and Pomegranate Peel Extracts* and *Natural Dyeing of Modified Cotton Fabric with Cochineal Dye*. The importance of leader research in the manuscript *Evaluation of DNA-Damaging Effects Induced by Different Tanning Agents Used in the Processing of Natural Leather Pilot Study on HepG2 Cell Line* should be particularly emphasised. The challenge of applying modern technologies in textile care is also included in the paper *Decontamination Efficiency of Thermal, Photothermal, Microwave, and Steam Treatments for Biocontaminated Household Textiles*. Increasing the efficiency of the household laundry process is also offered in the study *Influence of Hydrogen Peroxide on Disinfection and Soil Removal During Low Temperature Household Laundry*. The main environmental problem of the textile industry is the fact that its intensive development causes increased resource consumption, especially in terms of water consumption, and releases highly contaminated effluents. The benefits of natural adsorbents in textile waste water treatment are presented in the article *Influence of Initial pH value on the Adsorption of Reactive Black 5 Dye on Powdered Activated Carbon: Kinetics, Mechanisms, and Thermodynamics*.

Ana Sutlović, Sanja Ercegović Ražić, and Marija Gorjanc
Editors

Review

Mechanisms Involved in the Modification of Textiles by Non-Equilibrium Plasma Treatment

Gregor Primc, Rok Zaplotnik, Alenka Vesel and Miran Mozetič *

Department of Surface Engineering, Jožef Stefan Institute, 1000 Ljubljana, Slovenia
* Correspondence: miran.mozetic@ijs.si

Abstract: Plasma methods are often employed for the desired wettability and soaking properties of polymeric textiles, but the exact mechanisms involved in plasma–textile interactions are yet to be discovered. This review presents the fundamentals of plasma penetration into textiles and illustrates mechanisms that lead to the appropriate surface finish of fibers inside the textile. The crucial relations are provided, and the different concepts of low-pressure and atmospheric-pressure discharges useful for the modification of textile's properties are explained. The atmospheric-pressure plasma sustained in the form of numerous stochastical streamers will penetrate textiles of reasonable porosity, so the reactive species useful for the functionalization of fibers deep inside the textile will be created inside the textile. Low-pressure plasmas sustained at reasonable discharge power will not penetrate into the textile, so the depth of the modified textile is limited by the diffusion of reactive species. Since the charged particles neutralize on the textile surface, the neutral species will functionalize the fibers deep inside the textile when low-pressure plasma is chosen for the treatment of textiles.

Keywords: non-equilibrium plasma; textiles; surface modification

Citation: Primc, G.; Zaplotnik, R.; Vesel, A.; Mozetič, M. Mechanisms Involved in the Modification of Textiles by Non-Equilibrium Plasma Treatment. *Molecules* **2022**, *27*, 9064. https://doi.org/10.3390/molecules27249064

Academic Editors: Ana Sutlović, Sanja Ercegović Ražić and Marija Gorjanc

Received: 16 November 2022
Revised: 6 December 2022
Accepted: 7 December 2022
Published: 19 December 2022

Publisher's Note: MDPI stays neutral with regard to jurisdictional claims in published maps and institutional affiliations.

1. Introduction

Textiles are made from fibers of various materials, typically polymers. The properties of polymeric fibers depend on the type of materials and synthesis procedure. The ability of textiles to capture liquids depends on several parameters, and the most important is the surface wettability of the materials the fibers are made from. The surface wettability depends on the surface chemistry, particularly the polarity of surface functional groups. Most polymers contain an inadequate concentration of polar surface functional groups, so the wettability is usually below the level desired in numerous applications, particularly when attempting to bond chemically any foreign material. In cases when increased wettability is needed, the surface functional groups should be altered by grafting more polar groups. A common technique is the application of gaseous plasma. Although the technique has been used for some types of textiles on an industrial scale for decades, the scientific aspects are still a subject of investigation. Numerous scientific articles have been published on plasma modification of textiles, and the trend is shown in Figure 1. Only a few publications were published annually before 2000, and the number of publications increased by about 10-times in the past 20 years. Despite the increasing knowledge provided in these publications, the mechanisms involved in the modification of textile materials are still inadequately understood. In particular, the surface mechanisms on the atomic level are still a subject of research, even for smooth polymeric materials. The recommended theoretical papers on atomic-scale mechanisms on smooth polymer surfaces include [1–3], while the experimental observations on the surface kinetics were provided in [4,5]. Some review papers on plasma modification of textiles have also been published explaining the observations reported by various authors, such as [6–17]. Numerous books on various aspects of plasma treatment of textiles have been published, including a textbook intended for users of plasma technologies in the textile industry rather than experts in plasma science [18]. Despite a vast literature, the mechanisms of plasma-species penetration in fibrous materials, physical and chemical interactions with the plasma species, and the resulting surface finish of fibers deep inside the textiles are rarely explained to the level useful for scientists involved in the modification of woven and non-woven textiles with gaseous plasma. This paper intends to provide scientific aspects which are crucial for understanding plasma–textile interactions. Most recent papers are cited where appropriate. A reader will find prior publications in the citing literature.

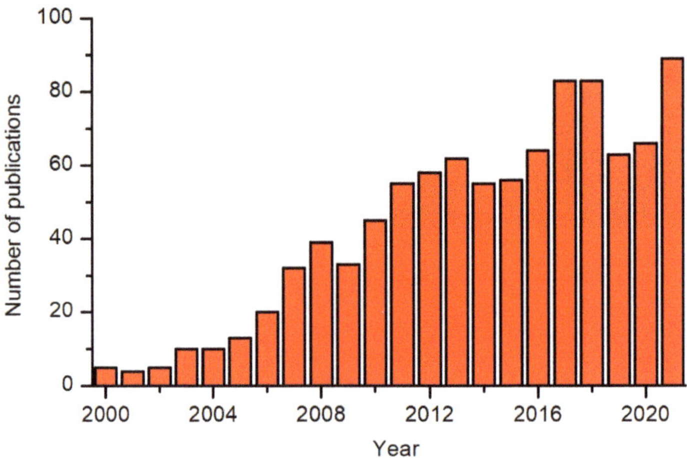

Figure 1. The number of publications published each year on plasma modification of textiles. The data were obtained from the Web of Science by searching with the keywords "textile" and "plasma" and "surface".

2. Gaseous Discharges and Plasma Species

Non-equilibrium plasma is usually sustained with an appropriate gaseous discharge. A voltage source is used to create an electric field strong enough to ensure the acceleration of free electrons, which thus gain appropriate energy for ionization collisions with neutral molecules. Various discharges have been reported by different authors. Low-pressure discharges usually operate with a power generator at high frequency in the range of radiofrequency (RF) or microwave (MW), roughly between 10 kHz and 10 GHz. The use of high-frequency fields is preferred since it ensures a rather homogeneous plasma in a large volume without risking arcing. Furthermore, high-frequency discharges can be operated in the electrodeless mode, i.e., any electrode placed outside the discharge chamber. This is useful because the electrodeless configuration overcomes all problems associated with an electrode placed inside the chamber, such as sputtering [19] the cathode by energetic ions and deposition of cathode material on treated samples, and the loss of reactive plasma species by heterogeneous surface recombination [20]. Namely, the loss of atoms on metallic surfaces is usually 10–1000-times more extensive than on surfaces of inert materials such as glasses and smooth polymers [21].

The microwave discharges (frequency between about 1 and 10 GHz) always operate in the electrodeless mode. The electric field provided by an MW generator will not penetrate deep into plasma because of the skin effect [22]. Namely, non-equilibrium gaseous plasma is a weakly or moderately conductive medium with an Ohmic character of the impedance, so the penetration depth of the electric field is limited to the skin layer between the dielectric discharge chamber and the gaseous plasma [23]. The higher the frequency, the thinner the skin layer. A textile sample is placed inside the plasma, so there is no electric field nearby except a small field due to the difference between the plasma and textile surface potentials, which creates a negative potential on the textile surface of about 10 V. As a consequence, no highly energetic gaseous specie will interact with the textile when an MW discharge is used.

The radiofrequency discharges operate either in the electrodeless mode or with an electrode mounted inside a metallic chamber, which is typically grounded [24]. The electrodeless mode is further divided into the E and H modes [25,26]. In the first case, the samples are usually at the floating potential, the same as if plasma is sustained with MW discharges. In the case of an electrode inside the chamber, the sputtering may be suppressed by careful selection of processing parameters but cannot be avoided, so at least a very thin metallic film (or clusters of metallic atoms) will deposit on the textile surface. Furthermore, extensive loss of molecular radicals will occur on the metallic surface, so the density of radicals, such as free atoms, will be orders of magnitude smaller than in cases plasma is sustained with an electrodeless RF discharge. Regardless of the excitation mechanisms, the penetration depth of the RF field inside plasma depends on plasma (not discharge) parameters and increases with decreasing frequency of the power source and with decreasing density of charged particles. This is a somewhat simplified illustration useful for scientists involved in the plasma treatment of textiles. Details are still a subject of scientific research [27,28].

As explained above, the charged particles are accelerated only in the volume of the significant electric field, which is always limited. In the rest of the discharge chamber, there will be diffusing plasma—plasma of practically zero electric field but rich in plasma species because of the presence of rather energetic electrons. The volume of a large electric field in a low-pressure gaseous plasma is often limited to a sheath next to the electrode, which may or may not be inserted in the discharge chamber. The loss of electrons' energy in the gas phase at low pressure is rather weak because of the low collision frequency, so practically all loss mechanisms take place on the surfaces. The same applies to other reactive plasma species (ions, atoms in the ground and metastable states, and metastable molecules). Typical examples of high-frequency low-pressure discharges and resulting plasmas are shown in Figure 2. Photos of inductively coupled RF plasma in the H and E modes are shown in Figure 3. A photo of a textile in a capacitive electrodeless plasma is shown in Figure 4.

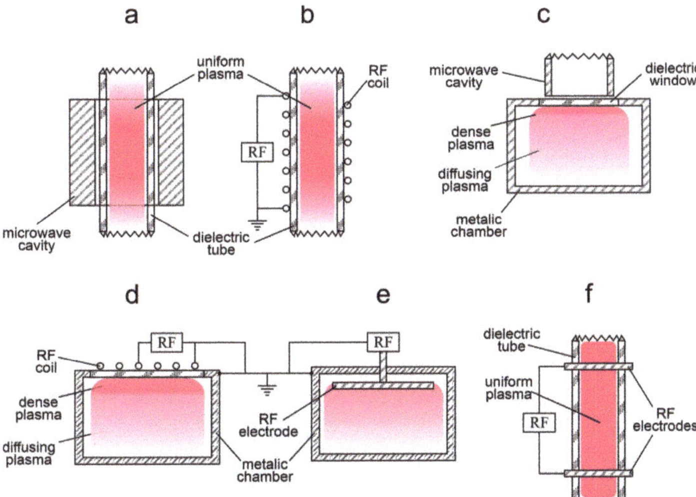

Figure 2. Examples of low-pressure high-frequency discharges. (**a**) Microwave in a dielectric tube, (**b**) inductive cylindrical RF in H-mode, (**c**) microwave in a metallic chamber, (**d**) inductive RF in a metallic chamber, (**e**) classical capacitive RF, and (**f**) electrodeless capacitive RF.

Figure 3. Photos of inductively coupled RF discharge in two distinguished modes: (**a**) H-mode (almost pure inductive character of the impedance) and (**b**) E-mode (with predominant capacitive component).

Figure 4. A photo of a textile in a plasma reactor powered with a rather low-power RF discharge.

Figure 3 shows the drastic difference between the intense plasma sustained by the RF discharge in the H-mode and the weakly ionized plasma sustained in the same tube and the same RF generator but in the E-mode. In fact, the plasma luminosity in the H mode is often several orders of magnitude larger than in the E-mode, and the reason is in excellent coupling between the generator and plasma electrons [29]. The electron density is much larger in the H-mode, but the density of neutral reactive species in both modes is comparable and much larger than the density of charged particles, usually above $1020 \ \mathrm{m}^{-3}$ [30]. The low-pressure plasma in the H-mode is, therefore, useful in cases when the material should be processed with charged particles and radiation but may not be suitable in cases when the neutral reactive species will do the job. Figures 3b and 4 also indicate a rather uniform plasma in a large volume when the discharge is coupled in the E-mode, while the H-mode RF plasmas are concentrated to a rather small volume, as also illustrated in Figure 1.

RF and MW discharges can also be used for sustaining plasma at atmospheric pressure, but the plasma volume will be limited to a small volume where the highest electric field appears. Namely, the diffusing plasma cannot expand to a large volume at atmospheric pressure due to numerous collisions of free electrons with gaseous molecules or atoms. The electrons lose their energy at inelastic (and, to a much lower extent, at elastic) collisions with molecules, so they do not diffuse far from the volume of a large electric field when the pressure is around the atmospheric (1 bar). Therefore, large spatial gradients in the density of charged particles are typical for atmospheric-pressure plasma sustained with high-frequency discharges, for example, at the industrial RF frequency of 13.56 MHz or MW frequency of 2.45 GHz. In addition, such discharges operate in the continuous mode even at atmospheric pressure, and the majority of discharge power is spent on heating the gas (increasing gas temperature), so the application of such high-frequency plasmas is limited at a pressure below, say, 30 mbar.

Low-frequency RF discharges or pulsed direct-current (DC) discharges are preferred plasma sources at atmospheric pressure. In such cases, plasma is not sustained in the continuous mode but rather in the form of stochastically distributed streamers [31]. The streamers are actually ionization wavefronts [32] born on an electrode and move away from the electrode at the speed of roughly 10^4 m/s [33,34]. They leave non-equilibrium gaseous plasma beyond them, but the plasma species are quickly thermalized at atmospheric pressure, so their density at a given position is marginal before the next streamer appears in the same volume unless plasma is sustained in high-purity noble gas [35]. Streamers are very narrow and short in duration, typically of the order of μm and μs, respectively. The maximal density of charged particles in a streamer is large, often above 10^{20} m^{-3} [36]. In the limiting case of repetitively pulsed discharges sustained by very fast switchers (switching time of the order of ns), the streamers form bullets that may propagate far from the electrode, sometimes close to 1 m [37].

Low-frequency and pulsed discharges at atmospheric pressure come in various modes. The scientific literature usually distinguishes two types of discharge: dielectric-barrier discharges (DBD) and corona discharges. DBD employs a dielectric barrier on an electrode to limit the current of a specific streamer, whereas corona employs a resistor mounted in series with the plasma. The electrical current in a particular streamer sustained with the DBD is limited by the available charge that could be transferred from the dielectric surface to the counter electrode. In the case of the corona, the voltage drop will appear on the resistor rather than on the gas gap when the electric current flows, so the voltage across the gas gap will drop below the value useful for sustaining the gaseous discharge during the current pulse. When the electric current diminishes, the entire available voltage (the source voltage) appears again across the gas gap, so another streamer appears. A schematic of the DBD and corona discharges are shown in Figure 5.

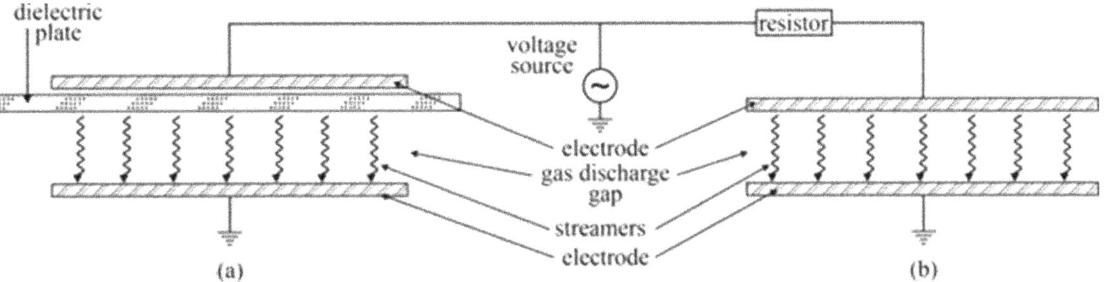

Figure 5. Schematic of the DBD (**a**) and the corona (**b**) discharges. The voltage source could be low-frequency RF, alternative current (AC), direct current (DC), or pulsed voltage source.

The illustrations in Figure 6 represent the basic principles of such discharges. In practice, there are various configurations. When atmospheric-pressure plasma is used, textiles are usually processed in the continuous mode to benefit from the automatization of the plasma processing. The speed of the textile running through plasma is often close to 1 m/s, so special configurations should be used. One of them is application of surface or coplanar discharges [38–40]. In the case of surface discharges, the electrodes are incorporated as conductive wires into a dielectric plate as shown schematically in Figure 6a. Streamers therefore conduct the electric current on the surface of the dielectric material. The reason for such surface streamers is that the breaking voltage for air at atmospheric pressure is much smaller than for the dielectric plate providing the plate is made from a material of very high resistance, for example, glass or some types of ceramics. The streamers will appear stochastically along wires, as in the classical DBD configuration shown in Figure 6, but will be dense so the entire surface of the dielectric plate will be covered by luminous plasma. Plasma will expand perhaps 1 mm from the surface, so the distance between the textile and the dielectric plate is a crucial parameter governing the surface finish of textiles treated with plasma sustained by surface discharges. A feasible solution is shown in Figure 6b. The system shown in Figure 6 operates well as long as the impedance of any material between the electrodes is much larger than the sum of impedances of dielectric material above the electrodes and gas above the dielectric plate (Figure 6a). This requirement is not trivial and dictates innovative solutions of coupling the power supply to the wires inside the dielectric plate. The connecting wires (not shown in Figure 6a) should be placed inside a perfectly insulating liquid that is cooled to prevent overheating. Namely, a significant fraction of the discharge power is spent on heating the dielectric plate.

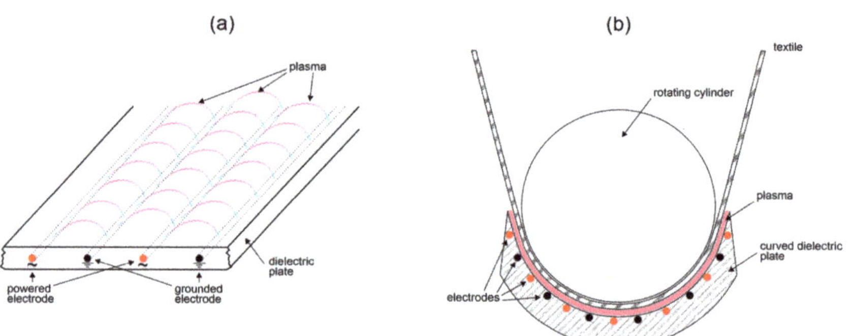

Figure 6. Schematic of the surface discharges sustained in the DBD mode (**a**) and a feasible solution to assure a constant distance between the textile and the dielectric plate for uniform treatment of textile surfaces (**b**).

Corona discharges useful for treatment in the continuous mode are usually sustained along one or more electrodes mounted next to a grounded roll. The textile runs in the space between the powered electrode and the grounded roll. The preferred embodiment is not much different from the one illustrated in Figure 6b, except that the corona discharge sustains plasma instead of the coplanar DBD. In both cases, it is vital to keep the distance between the powered electrodes and the textiles as constant as possible to enable uniform surface treatment.

Regardless of the type of discharge used for sustaining gaseous plasma, the following species are found in gaseous plasma:

- Free electrons and positively charged molecules or atoms;
- Neutral molecular radicals, including atoms, in the ground electronic state;
- Metastable atoms and molecules in excited electronic states;
- Radiation in the range from infrared (IR) through visible (vis) and ultraviolet (UV) to vacuum ultraviolet (VUV);
- Negatively charged ions (important in the cases plasma is sustained in electronegative gases).

The fluxes of these species on the surface of any samples, including textiles, depend enormously on the type of discharge used for sustaining gaseous plasma. Typical numerical values will be disclosed in the following text.

3. Penetration of Gaseous Plasma in Textiles

Usually, uniform treatment of fibers within a textile sample is preferred to the localized treatment of fibers stretching from a textile surface, so the penetration depth of species capable of modifying the fibers is important information. By definition, plasma is at least partially ionized gas with a density of charged particles large enough to form a space charge of linear dimension larger than the distance between two pieces of solid material (usually electrodes). The space charge is absolutely necessary for sustaining gaseous plasma; otherwise, the electrons would escape to the surfaces and will thus no longer be able to sustain plasma. The voltage is screened by the space charge over the distance with a characteristic thickness of the Debye length, which is defined as

$$\lambda_D = \sqrt{\frac{\varepsilon_0 k_B T_e}{e_0^2 n_e}}.$$

(1)

where T_e is the electron temperature, k_B is the Boltzmann constant, ε_0 is vacuum dielectric permeability ($\varepsilon_0 = 8.85 \times 10^{-12}$ F/m), e_0 is the elementary charge (1.6×10^{-19} As), and n_e is the electron density. The Debye length is proportional to the square root of the electron temperature. The electron temperature in gaseous plasma useful for the treatment of textiles does not span over a broad range; a typical value is between 10,000 and 100,000 K, which corresponds to the average electron energy of roughly between 1 and 10 eV. On the contrary, the Debye length is inversely proportional to the square root of the electron density, which depends significantly on the type of discharge and other discharge parameters and may be anywhere between about 10^{14} and 10^{20} m^{-3}. Obviously, the Debye length in plasma of electron temperature a few 10,000 K assumes any value between about 1 mm and 1 µm for low and high electron density, respectively. The Debye length versus the electron density is plotted in Figure 7a.

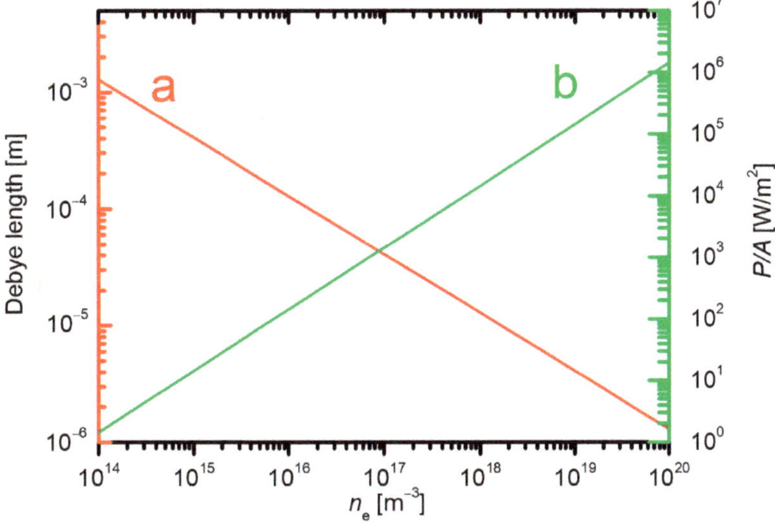

Figure 7. The Debye length (**a**) and the power dissipated on a surface facing a continuous plasma (**b**) versus the electron density at $kT_e = 3$ eV.

Low-pressure plasmas will have electron densities between about 10^{15} m^{-3} (this value is typical for diffusing plasma) and about 10^{18} m^{-3} (typical for powerful inductively coupled plasma in the H-mode or a microwave plasma sustained in a rather small discharge tube). The electron density of 10^{18} m^{-3} corresponds to the Debye length of 10 µm (Figure 7a). Obviously, the low-pressure plasma is not likely to penetrate deep into the textile unless a powerful discharge sustains it. A plasma of high density of charged particles will cause significant heating of an immersed material even if the material is kept at a floating potential. The geometrical surface of a plasma-facing material is subjected to heating by ions, which will neutralize on the surface and will bombard it with the kinetic energy gained by crossing the sheath between the non-perturbed plasma and the sample [41]. The power dissipated on the surface of any material (including textile) due to the interaction with positively charged ions is

$$P_{\text{charged}} = n_{\text{ion}} \cdot v_{\text{Bohm}} \cdot E_{\text{ion}} \cdot A. \tag{2}$$

where v_{Bohm} is Bohm velocity ($v_{\text{Bohm}} = \sqrt{k_{\text{B}} \cdot T_{\text{e}} / m_{\text{i}}}$), m_{i} is the mass of the positively charged ions, n_{ion} is positive ion density in the bulk plasma, E_{ion} is the sum of ionization energy and the kinetic energy of an ion, and A is the geometrical surface area. The dissipated power per surface area is plotted versus the electron density in Figure 7b. A plasma of the Debye length 10 μm (electron density 10^{18} m^{-3}) will cause heating at a power as large as 10^4 W/m^2. Obviously, either the density of charged power is too low to enable plasma penetration inside the textile, or it is too high to avoid overheating of the textile surface. From the plasma-penetration point of view, the low-pressure plasmas do not seem to be useful for the treatment of textiles unless the treatment time is extremely short. The benefits of using low-pressure plasmas will be explained below.

The density of charged particles within a streamer of an atmospheric pressure discharge is often much larger than in continuous plasma (for example, plasma sustained at low pressure). From this point of view, atmospheric-pressure plasma sustained by DBD or corona has advantages over continuous plasmas. The high plasma density will ensure for penetration of streamers in volume between the fibers. The schematic of the penetration of a streamer through the textile is illustrated in Figure 8. A streamer touches the textile surface (Figure 8a). The electrons are much faster than the ions (the velocity is $v = \sqrt{2W/m}$), where W is the kinetic energy and m the mass of a particle, so they charge the neighboring fibers negatively (Figure 8b). The negative surface charge will suppress the loss of electrons on surfaces, so the electron density within the streamer will remain large enough to sustain the plasma, which will propagate inside the textile (Figure 8c). Any obstacle on the streamer's way will be charged negatively (Figure 8d), thus re-directing further propagation (Figure 8e), so finally the streamer may actually penetrate the textile providing the electron density in the original streamer (Figure 8a) is large enough. The streamer will leave behind a partially dissociated gas, and the molecular radicals (in particular free atoms) will interact chemically with the fibers' surfaces. A benefit of such streamers propagating within porous material such as textile is that the molecular radicals are formed inside the textile (by electron-impact dissociation), so they can interact with fibers deep in the textile despite the short shelf-time typical for atmospheric-pressure plasmas. The drawback is non-uniform treatment because the streamers are stochastically distributed over the textile surface. The wettability, however, is a macroscopic parameter, so even a relatively small fraction of fibers inside the textile will ensure improved wettable and/or soaking properties. The effect is yet to be elaborated on in scientific literature, so the simplified explanation provided in this paragraph should be taken just as an illustration of the penetration of plasma streamers through textiles. The most important difference between streamers and continuous plasma is that the streamers will not destroy the textile by overheating because they carry negligible energy as compared to the flux of plasma species in the continuous mode.

The surface discharges powered by generators of frequency up to about a few 10 kHz illustrated in Figure 6 also consist of streamers, but those streamers will not penetrate the textile because they propagate on the dielectric's surface. The streamers of surface discharges will thus only touch the textile surface, which is beneficial for processing thin textiles but may not always be useful for reasonably uniform functionalization of thicker fabrics.

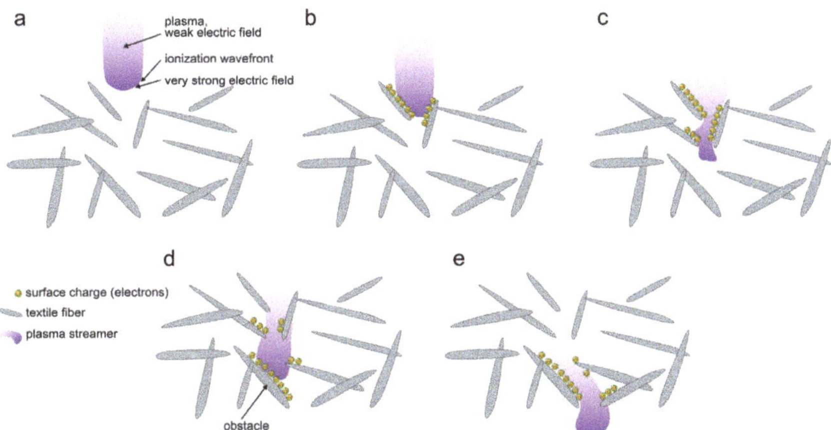

Figure 8. Illustration of a streamer penetration inside the textile and propagation through textile fibers from (**a**) to (**e**).

Differently to streamers, continuous plasma of reasonable density of charged particles will not penetrate the porous material because the Debye length is larger than the distance between neighboring fibers in a textile. Instead, the movement of reactive plasma particles will be governed by diffusion. One exception is the penetration of radiation. The interested photon energy is beyond the threshold for breaking bonds in polymer materials, i.e., above several eV, i.e., the UV and VUV range of wavelengths. The absorption depth of photons depends on the type of polymer and concentration of impurities that act as absorption sites. The absorption coefficient is a complex function of the photon energy even for pristine polymers [42], let alone textiles made from finite-purity materials. As a rule of thumb, the penetration depth decreases with increasing photon energy and becomes only 0.1 μm or less when the photon energy is above, say, 8 eV [43]. Such energetic radiation will, of course, penetrate the gaps between the fibers but will be absorbed into the relatively thin surface film of a fiber. If the textile porosity is large, the radiation will penetrate deep into the textile and break bonds in the surface film of polymer fibers. The dangling bonds may be occupied with gaseous molecules, so the functionalization with new functional groups may occur upon exposure of textiles to (V)UV radiation in the presence of a reactive gas such as oxygen.

The penetration depth of particles such as charged particles, metastables, and radicals in the textile treated with a continuous plasma will depend predominantly on the surface loss coefficients. As already mentioned and illustrated in Figure 8b, the electrons are fast as compared to any other plasma particle, so they will be the first to be lost by attachment to the fiber's surface. The positively charged ions will be attracted by the negative surface charge and will also be lost quickly because the surface neutralization efficiency is practically 100%. Negatively charged ions will be reflected upon entering the sheath, so they cannot reach the textile surface unless in cases of intentional RF biasing, which is beyond the scope of this text. On the other hand, metastables and radicals may suffer numerous collisions with the polymer surface before they finally relax, recombine, or are lost by chemical bonding to the polymer surface. The penetration depth of various plasma species in cases of streamer-free discharges (i.e., continuous plasma) is illustrated in Figure 9.

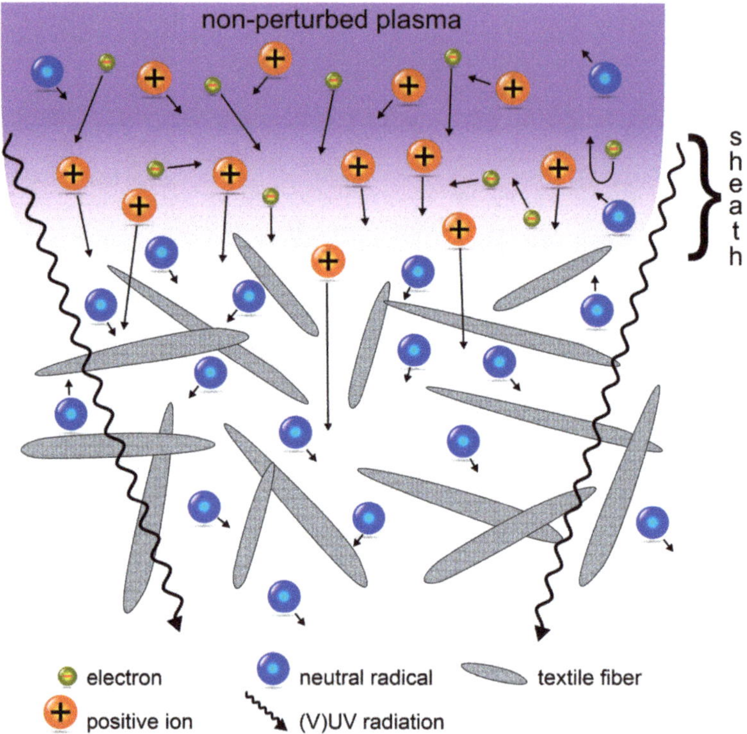

Figure 9. An illustration of the penetration path of VUV and UV radiation, charged particles, metastables, and radicals in textiles.

4. Surface Modifications Caused by Plasma Species

Once the penetration depths are known, it is possible to provide an illustration of the textile modifications upon treatment with gaseous plasma. Obviously, the inhomogeneous treatment over the entire thickness of the textile cannot be avoided but could be suppressed by the appropriate design of the treatment device and choice of the discharge parameters. Below are listed the effects of treatment by different plasma species.

4.1. Electrons

As illustrated in Figures 8 and 9, the electrons will cause a negative charge of the fibers on the surface (Figure 9) or even deep inside the textile (Figure 8). The negative charge will retard all but the fastest electrons arriving from the gaseous plasma toward the fibers' surfaces. As a result, the kinetic energy of a vast majority of electrons impinging the polymer surface will be below the threshold for breaking bonds. Therefore, the electrons have little effect on the textile modifications, except in the case of RF biasing, which is beyond the scope of this text. In almost all practical cases, the effect of electrons on the surface finish of textiles is marginal.

4.2. Positively Charged Ions

If the negatively charged electrons are retarded by the surface charge, the positively charged ions are accelerated toward the surface, as illustrated in Figures 8 and 9. There is always a successive negative charge on the surface of any material facing plasma, so the positively charged ions will impinge surfaces full of negatively charged electrons and will neutralize. Surface neutralization is very efficient, with the probability very close to 100%. The excessive energy (the ionization energy) will be dissipated on the fiber surface, so the surface will be heated. Furthermore, the kinetic energy of positively charged ions impinging the surface will also end up heating the fibers according to Equation (2). The kinetic energy of ions (about 10 eV in the most common case when the textile is left at floating potential) is larger than the bond strength in the polymer materials (several eV) so the ions impinging the surface are likely to interact chemically with the solid material. On the other hand, the kinetic energy of positively charged ions (roughly 10 eV when the textile is at floating potential) will not cause sputtering. The effect of the interaction between the positively charged ions and the pristine polymer is surface functionalization with groups that result from chemical interaction. If the ions are O_2^+, O^+, or $(OH)^+$, the newly formed functional groups will be C-OH, C=O, O-C=O, C-O-C, and so on. The effect will be limited to the very surface of the fiber unless some diffusion of successive surface oxygen occurs. The surface will soon be saturated with polar functional groups. Prolonged treatment will cause oversaturation, formation of low molecular weight fragments, and desorption of such fragments from the surface. Macroscopically, these effects will cause etching. The etching will be enhanced due to the heating caused by the dissipation of the potential and kinetic energy of positively charged ions impinging the surface. The etching is rarely laterally uniform, so nanostructuring of the fibers will occur. The combination of polar surface functional groups and rich morphology on the sub-micrometer scale will cause a super-hydrophilic surface finish [44] and, thus, excellent soaking dynamics upon wetting the textile. Obviously, the surface finish will depend on the fluence of ions on the surface, the surface temperature, and the type of polymer. The effect of positively charged ions is illustrated in Figure 10.

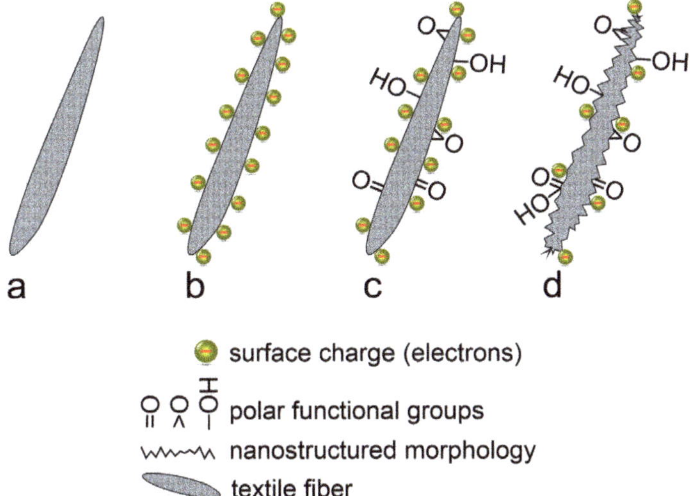

Figure 10. An illustration of a polymer fiber surface finishes upon interacting with positively charged oxygen ions: **a**—before interaction with plasma, **b**—immediately after turning on the plasma, **c**—after saturating the surface with polar groups, and **d**—after prolonged treatment.

4.3. Negatively Charged Ions

As already mentioned, the surfaces facing non-equilibrium gaseous plasma are biased negatively against the plasma because of the adsorption of free electrons from gaseous plasma. The negatively charged ions (relevant only for plasmas sustained in electronegative gases) will be retarded by the surface potential and be directed back to bulk plasma. Therefore, the negatively charged ions will not reach the fiber surface, so any discussion of interaction with the fibers is not necessary.

4.4. Metastables

Electrons of adequate energy will cause the excitation of both neutral molecules and atoms, as well as ions, into electronically excited states. The excitation occurs in the gaseous plasma where the electron density and temperature are rather large. The excited states may be resonant, meaning that they will immediately (in a time measured in nanoseconds) relax by emitting a photon. Many states, however, are metastable because of the rules of quantum physics, which prevent the relaxation by electrical dipole radiation. The lifetime of such excited states may be long enough to enable the metastables to reach the surface. On the surface, the excitation energy may be transferred to an electron bonded to the polymer surface, thus causing surface modification. The effect is yet to be elaborated because very few results on the interaction between oxygen metastables and polymer materials have been published, although plasma scientists often take them into account when explaining mechanisms in oxygen plasmas [45–47].

4.5. Molecular Radicals, including Atoms

The neutral reactive particles usually govern the surface chemistry upon treatment of polymer materials with oxygen plasma [48]. In almost all practical cases, their concentration in gaseous plasma is orders of magnitude larger than the concentration of charged particles [20]. In low-pressure plasmas, the main reason for the large concentration of neutral reactive particles is the excellent stability in the gas phase. Namely, the recombination of simple radicals like atoms to parent molecules in the gas phase requires a three-body collision. Such collisions are scarce at pressures below a few mbar. The lifetime of atoms in a properly designed experimental system is more than a millisecond [49], so the radicals may pass several meters from the source (i.e., dense plasma) [50]. The recombination in low-pressure systems takes place practically exclusively on surfaces. Unlike the neutralization of charged particles with almost 100% probability, the surface recombination of neutral radicals depends enormously on the type of materials facing plasma [51], and may be as low as 0.001 for some types of polymers [52]. The neutral reactive species will therefore diffuse in the space between the fibers of the textile and may actually cause significant chemical modification of fibers deep inside the textile, providing the loss on polymer fibers by surface recombination is marginal. The atoms are often in thermal equilibrium with the neutral gas, so their temperature will not be much over room temperature. As a consequence, the atoms will not cause heating of the textile unless a chemical or physical reaction occurs on the fiber surface. The physical reaction is surface recombination to parent molecules, while the chemical reactions cause polymer oxidation. The oxidation kinetics are complex. For example, Longo et al. [1] found as many as 20 binding sites with different adsorption energies even for pristine polymer, let alone partially oxidized material. By far the most probable interaction is the substitution of C-H with C-OH [1]. Hydroxyl groups are, therefore, the first to be formed on a polymer surface. In fact, experiments show that the surface saturation with hydroxyl groups occurs at the dose of O atoms of 10^{19} m^{-2} [5]. If the atom density next to the fiber is 10^{21} m^{-3}, the saturation occurs in 0.1 ms! Much larger doses are needed for saturating the surface with other oxygen-containing functional groups, but 10^{22} m^{-2} is enough for many polymers, including fluorinated ones [4].

The unique advantage of using neutral radicals over positively charged ions is excellent efficiency for surface oxidation. An atom will either interact chemically with a polymer surface or will experience an elastic collision and will not cause surface heating. The heating also occurs at surface recombination, but the coefficient is low for most polymers [52], so the atoms represent the plasma species of choice when functionalization of fibers deep in the textile is desired without significant heating of the textile surface. The interaction of O atoms with a polymer fiber is shown in Figure 11. Large doses will cause etching, but the effect is marginal as compared to etching rates observed for treatment with positively charged ions.

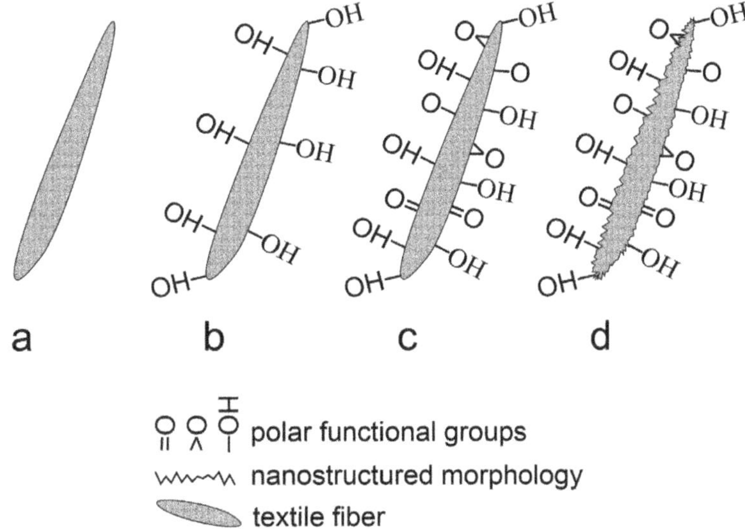

Figure 11. An illustration of a polymer fiber surface finishes upon interacting with neutral oxygen atoms. **a**—before interaction, **b**—after receiving the dose of about 10^{19} m^{-3}, **c**—after receiving the dose of about 10^{22} m^{-3}, and **d**—after very large doses.

At atmospheric pressure, the loss of radicals in the gas phase is efficient in molecular gases and less efficient in noble gases with a small admixture (typically up to about 1 vol.%) of molecular gas. That is one of the reasons for using noble gases in atmospheric-pressure plasmas. In any case, the maximal achievable atom density at atmospheric plasmas is practically the same as in low-pressure plasmas, i.e., roughly about 10^{21} m^{-3} [53]. The energy efficiency of atmospheric-pressure plasmas, especially those sustained in molecular gases such as oxygen or air, is inferior to the efficiency of low-pressure plasmas sustained in dielectric tubes by electrodeless discharges because of the extensive recombination of atoms to parent molecules at three-body collisions at atmospheric pressure.

5. Conclusions

An insight into the plasma techniques for the functionalization of textiles from the perspective of plasma science was presented. Plasma sustained in pure noble gases does not contain chemically reactive particles but is a source of VUV and UV radiation which breaks bonds in the polymer material and thus provides dangling bonds to be occupied upon exposure to reactive gases. The efficiency is rather inadequate because the photons are not absorbed in the very-surface film useful for functionalization with desired surface functional groups but rather in a thicker film, typically about 100 nm for very energetic photons of 8 eV and above, more for less energetic photons. The effect of other plasma species is limited to the very surface, especially when the textile is kept at a floating potential (which is true in all practical cases). The reactive plasma particles are predominantly formed at an impact with an energetic electron in gaseous plasma, but the loss will be in the gas phase at elevated pressure (atmospheric and above), and on surfaces at low pressure (a few mbar and below). In between atmospheric and low pressure, there is a range of pressures rarely tackled by scientists or users of plasma technologies. The energy dissipated at the neutralization of charged particles and the recombination of radicals, including atoms, to stable molecules will cause heating. Gaseous molecules will be heated in atmospheric-pressure plasmas, while in low pressure, the surfaces will be heated. Low-pressure plasmas are usually sustained in the continuous mode and are rich in neutral reactive particles that are useful for chemical interaction with polymer surfaces. While the atmospheric-pressure plasmas can be sustained in the continuous mode, they are not very useful for the treatment of textiles due to the high-power density needed for sustaining the continuous plasma at high pressure. From this aspect, it is advisable to use atmospheric plasmas, which operate in streamers—current pulses of duration of the order of µs occurring stochastically in the discharge gap. The streamers are actually ionization wavefronts progressing from the powered electrode at a speed of roughly 10^4 m/s. The ionization wavefronts leave behind non-equilibrium gaseous plasma, which is of short duration because of numerous collisions which lead to the gas thermalization and thus the loss of chemically reactive particles useful for surface functionalization of polymers. An advantage that makes the atmospheric-pressure plasmas attractive for the treatment of textiles is the ability to sustain plasma even inside the textile, in the space between the fibers, providing the electron density within the streamer is large enough so that the space charge can suppress their loss on the surfaces. By using streamers of high electron density, it is possible to obtain the desired surface finish of the entire textile despite the localized character of the streamers. On the contrary, low-pressure plasmas assure a large density of molecular radicals, in particular atoms, which may diffuse in the space between the fibers and cause surface chemistry even on fibers rather deep in the textiles. The charged particles will always cause significant heating of the materials facing them, so the plasma of a large density of charged particles may cause overheating of the fibers on the textile surface while leaving the fibers deep inside the textile inadequately treated.

Author Contributions: Conceptualization, M.M.; methodology, G.P.; writing—original draft preparation, G.P., R.Z., A.V., and M.M.; writing—review and editing, R.Z. and M.M.; visualization, G.P.; supervision, M.M.; project administration, A.V.; funding acquisition, M.M. All authors have read and agreed to the published version of the manuscript.

Funding: This research was funded by Slovenian Research Agency, grant number L2-2616, and core fund P2-0082.

Institutional Review Board Statement: Not applicable.

Informed Consent Statement: Not applicable.

Data Availability Statement: Not applicable.

Acknowledgments: The authors acknowledge the financial support from the Slovenian Research Agency, grant number L2-2616 (Selective surface modification of polymers), and core fund P2-0082 (Thin film structures and plasma-surface engineering).

Conflicts of Interest: The authors declare no conflict of interest.

Sample Availability: Samples of the compounds are available from the authors.

References

1. Longo, R.C.; Ranjan, A.; Ventzek, P.L.G. Density Functional Theory Study of Oxygen Adsorption on Polymer Surfaces for Atomic-Layer Etching: Implications for Semiconductor Device Fabrication. *ACS Appl. Nano Mater.* **2020**, *3*, 5189–5202. [CrossRef]
2. Fukunaga, Y.; Longo, R.C.; Ventzek, P.L.G.; Lane, B.; Ranjan, A.; Hwang, G.S.; Hartmann, G.; Tsutsumi, T.; Ishikawa, K.; Kondo, H.; et al. Interaction of oxygen with polystyrene and polyethylene polymer films: A mechanistic study. *J. Appl. Phys.* **2020**, *127*, 023303. [CrossRef]
3. Polito, J.; Denning, M.; Stewart, R.; Frost, D.; Kushner, M.J. Atmospheric pressure plasma functionalization of polystyrene. *J. Vac. Sci. Technol. A* **2022**, *40*, 043001. [CrossRef]
4. Lojen, D.; Zaplotnik, R.; Primc, G.; Mozetič, M.; Vesel, A. Optimization of surface wettability of polytetrafluoroethylene (PTFE) by precise dosing of oxygen atoms. *Appl. Surf. Sci.* **2022**, *598*, 153817. [CrossRef]
5. Vesel, A.; Zaplotnik, R.; Mozetič, M.; Primc, G. Surface modification of PS polymer by oxygen-atom treatment from remote plasma: Initial kinetics of functional groups formation. *Appl. Surf. Sci.* **2021**, *561*, 150058. [CrossRef]
6. Buyle, G. Nanoscale finishing of textiles via plasma treatment. *Mater. Techn.* **2013**, *24*, 46–51. [CrossRef]
7. Morent, R.; De Geyter, N.; Verschuren, J.; De Clerck, K.; Kiekens, P.; Leys, C. Non-thermal plasma treatment of textiles. *Surf. Coat. Technol.* **2008**, *202*, 3427–3449. [CrossRef]
8. Tessier, D. Surface modification of biotextiles for medical applications. In *Biotextiles as Medical Implants*; King, M.W., Gupta, B.S., Guidoin, R., Eds.; Woodhead Publishing Ltd.: Cambridge, UK, 2013; pp. 137–156. [CrossRef]

9. Saleem, M.; Naz, M.Y.; Shoukat, B.; Shukrullah, S.; Hussain, Z. Functionality and applications of non-thermal plasma activated textiles: A review. *Mater. Today Proc.* **2021**, *47*, S74–S82. [CrossRef]

10. Verschuren, J.; Kiekens, P.; Leys, C. Textile-specific Properties that Influence Plasma Treatment, Effect Creation and Effect Characterization. *Text. Res. J.* **2016**, *77*, 727–733. [CrossRef]

11. Zille, A.; Oliveira, F.R.; Souto, A.P. Plasma Treatment in Textile Industry. *Plasma. Process. Polym.* **2015**, *12*, 98–131. [CrossRef]

12. Jelil, R.A. A review of low-temperature plasma treatment of textile materials. *J. Mater. Sci.* **2015**, *50*, 5913–5943. [CrossRef]

13. Guimond, S.; Hanselmann, B.; Amberg, M.; Hegemann, D. Plasma functionalization of textiles: Specifics and possibilities. *Pure Appl. Chem.* **2010**, *82*, 1239–1245. [CrossRef]

14. Peran, J.; Ercegović Ražić, S. Application of atmospheric pressure plasma technology for textile surface modification. *Text. Res. J.* **2019**, *90*, 1174–1197. [CrossRef]

15. Tudoran, C.; Roşu, M.C.; Coroş, M. A concise overview on plasma treatment for application on textile and leather materials. *Plasma. Processes. Polym.* **2020**, *17*, 2000046. [CrossRef]

16. Mowafi, S.; Abou Taleb, M.; El-Sayed, H.E.-D. A Review of Plasma-assisted Treatments of Textiles for Eco-friendlier Water-less Processing. *Egypt. J. Chem.* **2022**, *5*, 737–749. [CrossRef]

17. Yilma, B.B.; Luebben, J.F.; Nalankilli, G. Cold Plasma Treatment in Wet Chemical Textile Processing. *Fibres Text. East. Eur.* **2020**, *28*, 118–126. [CrossRef]

18. Gorjanc, M.; Mozetic, M. *Modification of Fibrous Polymers by Gaseous Plasma*; LAP LAMBERT Academic Publishing: Saartbrücken, Germany, 2014.

19. Petrovska, S.; Sergiienko, R.; Ilkiv, B.; Nakamura, T.; Ohtsuka, M. Influence of sputtering power on optical, electrical properties and structure of aluminum-doped indium saving indium-tin oxide thin films sputtered on preheated substrates. *Mol. Cryst. Liquid Cryst.* **2022**, 1–9. [CrossRef]

20. Primc, G. Generation of Neutral Chemically Reactive Species in Low-Pressure Plasma. *Front. Phys.* **2022**, *10*, 895264. [CrossRef]

21. Wickramanayaka, S.; Meikle, S.; Kobayashi, T.; Hosokawa, N.; Hatanaka, Y. Measurements of catalytic efficiency of surfaces for the removal of atomic oxygen using NO2* continuum. *J. Vac. Sci. Technol. A-Vac. Surf. Films* **1991**, *9*, 2999–3002. [CrossRef]

22. Lee, H.-C.; Oh, S.; Chung, C.-W. Experimental observation of the skin effect on plasma uniformity in inductively coupled plasmas with a radio frequency bias. *Plasma Sources Sci. Technol.* **2012**, *21*, 035003. [CrossRef]

23. Apostol, M. Penetration depth of an electric field in a semi-infinite classical plasma. *Optik* **2020**, *220*, 165009. [CrossRef]

24. Chabert, P.; Tsankov, T.V.; Czarnetzki, U. Foundations of capacitive and inductive radio-frequency discharges. *Plasma Sources Sci. Technol.* **2021**, *30*, 024001. [CrossRef]

25. Zaka-ul-Islam, M.; O'Connell, D.; Graham, W.G.; Gans, T. Electron dynamics and frequency coupling in a radio-frequency capacitively biased planar coil inductively coupled plasma system. *Plasma Sources Sci. Technol.* **2015**, *24*, 044007. [CrossRef]

26. Lojen, D.; Zaplotnik, R.; MozetiČ, M.; Vesel, A.; Primc, G. Power characteristics of multiple inductively coupled RF discharges inside a metallic chamber. *Plasma Sci. Technol.* **2021**, *24*, 015403. [CrossRef]

27. Tsankov, T.V.; Chabert, P.; Czarnetzki, U. Foundations of magnetized radio-frequency discharges. *Plasma Sources Sci. Technol.* **2022**, *31*, 084007. [CrossRef]

28. Otsuka, F.; Hada, T.; Shinohara, S.; Tanikawa, T. Penetration of a radio frequency electromagnetic field into a magnetized plasma: One-dimensional PIC simulation studies. *EPS* **2015**, *67*, 85. [CrossRef]

29. Drašković-Bračun, A.; Mozetič, M.; Zaplotnik, R. E- and H-mode transition in a low pressure inductively coupled ammonia plasma. *Plasma. Processes. Polym.* **2018**, *15*, 1700105. [CrossRef]

30. Zaplotnik, R.; Vesel, A.; Mozetic, M. Transition from E to H mode in inductively coupled oxygen plasma: Hysteresis and the behaviour of oxygen atom density. *EPL* **2011**, *95*, 55001. [CrossRef]

31. Tarasenko, V.F.; Naidis, G.V.; Beloplotov, D.V.; Lomaev, M.I.; Sorokin, D.A.; Babaeva, N.Y. Streamer Breakdown of Atmospheric-Pressure Air in a Non-Uniform Electric Field at High Overvoltages. *Russ. Phys. J.* **2018**, *61*, 1135–1142. [CrossRef]

32. Huang, B.; Zhang, C.; Zhu, W.; Lu, X.; Shao, T. Ionization waves in nanosecond pulsed atmospheric pressure plasma jets in argon. *High Volt.* **2021**, *6*, 665–673. [CrossRef]

33. Wang, R.; Xu, H.; Zhao, Y.; Zhu, W.; Zhang, C.; Shao, T. Spatial–Temporal Evolution of a Radial Plasma Jet Array and Its Interaction with Material. *Plasma Chem. Plasma Process.* **2018**, *39*, 187–203. [CrossRef]

34. Hofmans, M.; Viegas, P.; Rooij, O.v.; Klarenaar, B.; Guaitella, O.; Bourdon, A.; Sobota, A. Characterization of a kHz atmospheric pressure plasma jet: Comparison of discharge propagation parameters in experiments and simulations without target. *Plasma Sources Sci. Technol.* **2020**, *29*, 034003. [CrossRef]

35. Li, J.; Lei, B.; Wang, J.; Xu, B.; Ran, S.; Wang, Y.; Zhang, T.; Tang, J.; Zhao, W.; Duan, Y. Atmospheric diffuse plasma jet formation from positive-pseudo-streamer and negative pulseless glow discharges. *Commun. Phys.* **2021**, *4*, 64. [CrossRef]

36. Jiang, C.; Miles, J.; Hornef, J.; Carter, C.; Adams, S. Electron densities and temperatures of an atmospheric-pressure nanosecond pulsed helium plasma jet in air. *Plasma Sources Sci. Technol.* **2019**, *28*, 085009. [CrossRef]

37. Schweigert, I.; Zakrevsky, D.; Gugin, P.; Yelak, E.; Golubitskaya, E.; Troitskaya, O.; Koval, O. Interaction of Cold Atmospheric Argon and Helium Plasma Jets with Bio-Target with Grounded Substrate Beneath. *Appl. Sci.* **2019**, *9*, 4528. [CrossRef]

38. Štěpánová, V.; Šrámková, P.; Sihelník, S.; Stupavská, M.; Jurmanová, J.; Kováčik, D. The effect of ambient air plasma generated by coplanar and volume dielectric barrier discharge on the surface characteristics of polyamide foils. *Vacuum* **2021**, *183*, 109887. [CrossRef]

39. Homola, T.; Kelar, J.; Černák, M.; Kováčik, D. Large-area open air plasma sources for roll-to-roll manufacture. *VIP* **2022**, *34*, 21–25. [CrossRef]

40. Sramkova, P.; Kelar Tucekova, Z.; Fleischer, M.; Kelar, J.; Kovacik, D. Changes in Surface Characteristics of BOPP Foil after Treatment by Ambient Air Plasma Generated by Coplanar and Volume Dielectric Barrier Discharge. *Polymers* **2021**, *13*, 4173. [CrossRef]

41. Booth, J.-P.; Mozetič, M.; Nikiforov, A.; Oehr, C. Foundations of plasma surface functionalization of polymers for industrial and biological applications. *Plasma Sources Sci. Technol.* **2022**, *31*, 103001. [CrossRef]

42. Zhang, Y.; Kong, L.; Du, H.; Zhao, J.; Xie, Y. Three novel donor-acceptor type electrochromic polymers containing 2,3-bis(5-methylfuran-2-yl)thieno[3,4-b]pyrazine acceptor and different thiophene donors: Low-band-gap, neutral green-colored, fast-switching materials. *J. Electroanal. Chem.* **2018**, *830–831*, 7–19. [CrossRef]

43. Fouchier, M.; Pargon, E.; Azarnouche, L.; Menguelti, K.; Joubert, O.; Cardolaccia, T.; Bae, Y.C. Vacuum ultra violet absorption spectroscopy of 193 nm photoresists. *Appl. Phys. A* **2011**, *105*, 399–405. [CrossRef]

44. Mozetic, M. Plasma-Stimulated Super-Hydrophilic Surface Finish of Polymers. *Polymers* **2020**, *12*, 2498. [CrossRef] [PubMed]

45. Gudmundsson, J.T. Recombination and detachment in oxygen discharges: The role of metastable oxygen molecules. *J. Phys. D-Appl. Phys.* **2004**, *37*, 2073–2081. [CrossRef]

46. Gudmundsson, J.T.; Hannesdóttir, H. On the role of metastable states in low pressure oxygen discharges. *AIP Conf. Proc.* **2017**, *1811*, 120001.

47. Sousa, J.S.; Niemi, K.; Cox, L.J.; Algwari, Q.T.; Gans, T.; O'Connell, D. Cold atmospheric pressure plasma jets as sources of singlet delta oxygen for biomedical applications. *J. Appl. Phys.* **2011**, *109*, 123302. [CrossRef]

48. Vesel, A.; Primc, G.; Zaplotnik, R.; Mozetič, M. Applications of highly non-equilibrium low-pressure oxygen plasma for treatment of polymers and polymer composites on an industrial scale. *Plasma Phys. Control. Fusion* **2020**, *62*, 024008. [CrossRef]

49. Ricard, A.; Gaillard, M.; Monna, V.; Vesel, A.; Mozetic, M. Excited species in H2, N2, O2 microwave flowing discharges and post-discharges. *Surf. Coat. Technol.* **2001**, *142–144*, 333–336. [CrossRef]

50. Zaplotnik, R.; Vesel, A.; Mozetic, M. A Powerful Remote Source of O Atoms for the Removal of Hydrogenated Carbon Deposits. *J. Fusion Energ.* **2012**, *32*, 78–87. [CrossRef]

51. Dickens, P.G.; Sutcliffe, M.B. Recombination of oxygen atoms on oxide surfaces. Part 1.—Activation energies of recombination. *Trans. Faraday Soc.* **1964**, *60*, 1272–1285. [CrossRef]

52. Kristof, J.; Macko, P.; Veis, P. Surface loss probability of atomic oxygen. *Vacuum* **2012**, *86*, 614–619. [CrossRef]

53. Reuter, S.; von Woedtke, T.; Weltmann, K.D. The kINPen-a review on physics and chemistry of the atmospheric pressure plasma jet and its applications. *J. Phys. D-Appl. Phys.* **2018**, *51*, 233001. [CrossRef]

Article

Influence of Cotton Cationization on Pigment Layer Characteristics in Digital Printing

Martinia Ira Glogar, Tihana Dekanić *, Anita Tarbuk, Ivana Čorak and Petra Labazan

Department of Textile Chemistry and Ecology, Faculty of Textile Technology, University of Zagreb, HR-10000 Zagreb, Croatia; martinia.glogar@ttf.unizg.hr (M.I.G.); anita.tarbuk@ttf.unizg.hr (A.T.); ivana.corak@ttf.unizg.hr (I.Č.); 5ra.labazan@gmail.com (P.L.)
* Correspondence: tihana.dekanic@ttf.unizg.hr

Abstract: The paper examines the influence of cotton cationization on the print quality in terms of penetration, colour yield and colour depth, which have been analysed in comparison to cotton untreated and pretreated with conventional acrylate binder. The process of cationization during mercerization was performed with a cationizing agent Rewin DWR (CHT Bezema). Standard (non-cationized) and cationized fabric, with and without additional layering of binder have been printed by digital inkjet pigment printing method. Moisture management testing (MMT) and dynamic contact angle measurement (drop shape analyzer–DSA30S) were performed on standard and cationized fabric, with and without binder, both with and without pigment layer. After printing, the objective values of colour depth (K/S) and colour parameters L*, C* and h° were analysed. The samples were also analysed by the method of microscopic imaging using a DinoLite microscope. Printed samples were tested to washing fastness, and the results are presented in terms of total colour difference (dECMC), according to CMC(l:c) equation, after the 1st, 3rd, 5th, 7th and 10th washing cycles. Results showed that the cotton cationization will improve the uniformity and coverage of the printed area as well as increase the K/S value. For the samples with binder, the positive effect of cationization on the stability and bond strength between the polymer layer as a pigment carrier with the cotton fabric was confirmed.

Keywords: inkjet printing; cationization; cotton; pigment; wash fastness

Citation: Glogar, M.I.; Dekanić, T.; Tarbuk, A.; Čorak, I.; Labazan, P. Influence of Cotton Cationization on Pigment Layer Characteristics in Digital Printing. *Molecules* **2022**, *27*, 1418. https://doi.org/10.3390/molecules27041418

Academic Editor: Giulio Malucelli

Received: 30 December 2021
Accepted: 17 February 2022
Published: 19 February 2022

Publisher's Note: MDPI stays neutral with regard to jurisdictional claims in published maps and institutional affiliations.

1. Introduction

Despite its unquestionable positive aspects, digital printing on textiles has not yet experienced its full commercialization and significant industrial application. According to a survey conducted by Zimmer in 2017, 97% of total world production of printed textiles is still carried out by conventional rotary printing technology [1]. The reasons for this are multiple and highly complex, arising from several contexts related to digital printing key issues: substrate characteristics, printing ink formulation, choice of dye or pigment, mechanism of interaction of printing ink droplets and textile substrate, technical requirements of machines and especially print heads for digital printing and requirements on the rheological properties of printing inks for digital printing. Currently, the most important directions of research in the field of digital textile printing is in innovative approaches in the printing ink formulation and modifications and pre-treatments of textile materials. In the application of pigments in printing ink formulation for digital textile printing, requests pertaining to pigment particle size, restricted content of binders and crosslinking agents in printing inks as well as their requested very low viscosity with higher surface tension opens an extensive platform for researching innovative formulations of binders, as well as innovative methods of textile surface pre-treatment [2–4]. Xue et al. experimented with the polymeric binders based on nanoscale emulsions containing different ratios of soft and hard monomers, coming out with the optimal results by applying a mixture of butyl acrylate soft monomers and methyl methacrylate hard monomers in ratio 8:1, employing, N-methylolacrylamide as crosslinking agents and monododecyl maleate as copolymerizable surfactant [5]. In the context of nanoscale binders, also, extensive research is being conducted in the field of chitosan application, as a pre-treating biomaterial [6,7]. As for the structural and chemical modification of a textile surface, in order to achieve optimal pigment layering and colour yield, with satisfactory fastness properties, plasma pre-treatment and cationic compounds have been used [8]. A review of the literature confirms the extensive research work in the field of application of cotton pre-treatment by cationization in digital printing processes based on dyestuff printing ink formulations [9–11]. Yang et al. [12] and Rekaby et al. [13] reported improved colour strength, increased colour yield, better pattern sharpness, higher colourfastness and reduced the steaming duration as well as washing-off procedure in reactive inkjet printing on cotton pre-treated by cationization. As for the cationic pre-treatment in pigment-based inkjet printing, the number of published studies is rather lower. Hauser and Kanik et al. studied printing of cationized cotton using reactive, direct and acid dye as well as pigments [14–18]. Wang and Zhang applied synergy of both cotton modification by commercially available cationic compounds Cibafix Eco and pigment modification, particularly the size of a pigment, achieving better colour strength and lower pigment penetration [19–21]. El-Shishtawy studied anionic dye and pigment printing. Cationization was applied as pre-treatment or after-treatment by exhaustion or padding and contributed to the better colour yield and fastness in all cases [22]. Grancarić, Tarbuk and Dekanić introduced cationization during mercerization, which results in cellulose with properties of both processes [23,24]. Application of the commercial long-chain cationic compounds in pre-treatment or after-treatment by exhaustion or padding, results in blocking of the surface active groups of cellulose, and if the dyeing process is performed afterwards, the coloration usually is not uniform. On the other hand, if the cationization is performed during the mercerization process, new cellulose is formed. Change in cellulose crystal lattice occurs and at the same time the cationic compound us evenly distributed and trapped between the cellulose chains, resulting in levelness of colour. There are no published papers related to printing on such modified cotton cellulose, especially ink-jet pigment printing and few papers related to dyeing with direct, acid, natural, vat and reactive dyes within same group of authors [25–30]. Therefore, the aim of this research was to investigate and analyse the pigment prints on samples of standard cotton fabric, untreated and modified by the cationization during the mercerization process considering binding of pigment ink and binder itself.

The aim of this work was to study the behaviour of the pigment-based inkjet print in dependence on the changes of substrate characteristics. The analyses were performed from the aspect of textile substrate pre-treatment, without interfering with constant and industrially defined settings of the printing process. Samples were digitally printed and the coverage of the textile surface with pigment, the uniformity of the pigment layer, the colour depth and objective colour parameters were examined. The wash fastness test was also performed and, in addition to objective evaluation, microscopic imaging of printed surfaces before and after washing was performed in order to determine the level of damage to the pigment polymer layer in the washing process, depending on fabric pre-treatment.

2. Materials and Methods

Standard fabric (WFK Switzerland, type 10A for ISO 2267) of 100% cotton was chosen, in canvas embroidery P1/1 with the following physical and mechanical characteristics:

- Standard, chemically bleached fabric, a flat weight of $170 \, \text{g/m}^2$, the density of the warp and weft was 27/27 threads per cm, and the fineness of the yarn was 295 dtex/295 dtex;
- Cationized cotton fabric during mercerization with Rewin DWR: mass per unit area of $185 \, \text{g/m}^2$ was measured, and the density of the warp and weft is 29/28.5 threads per cm.

The process of cationization during mercerization was performed according to [24] with a cationizing agent, Rewin DWR (CHT-Bezema, Montlingen, Switzerland). It was carried out at a continuous rate of v = 3 m/min, with 0% tenacity, at 20 °C, with 5 passes through the bath on a Jigger machine. First, the mercerization of 5 passages in 24% NaOH and 8 g/L of Subitol MLF (CHT-Bezema, Montlingen, Switzerland) was performed, followed by 5 passages in 50 g/L of Rewin DWR. Afterward, cotton fabric crosslinking occurs in a closed system for 24 h, followed by hot rinsing (fixation) with distilled water at 100 °C for 2 min. The cotton fabric is further rinsed, neutralized with 5% acetic acid (CH_3COOH), followed by cold rinsing to achieve a neutral pH, and air-dried.

2.1. Fabric Surface Characterization

For the characterization of the changes of interface phenomena of the cotton fabric, zeta potential, absorption rate and contact angle were determined.

Electrokinetic potential (zeta, ζ) was measured using an electrokinetic analyser, EKA (Anton Paar, Graz, Austria). Measurement was performed according to the following conditions: sample mass = 0.2000 g; V_{KCl} = 500 mL; c_{KCl} = 10 − 3 mol/L; d = 0.55 mm (electrode spacing); p = 300 mbar; pH 2.5 to 10. The results of the measured zeta potential were calculated according to the Helmholtz-Smoluchowsky equation [30].

The moisture management test was performed according to AATCC 195-2020 *Liquid Moisture Management Properties of Textile Fabrics* on a moisture management tester (MMT). For testing, a solution prepared according to the producer specification was used: 9 g/L of NaCl on 1 L of distilled water, with the conductivity of 16 mS/ ± 0.2 mS. From measured properties, mean values of absorption rate (AR) and total ability to manage (liquid) moisture (OMMC) are presented, and type of fabric is given.

Measurements of the dynamic contact angle on the textile material surface were performed on a goniometer (Drop Shape Analyzer- DSA30S, Kruess, Germany) with the distilled water as test liquid.

2.2. Inkjet Printing

The printing of cotton samples was performed by digital inkjet technique on an Azon Tex Pro textile inkjet printer (Azon d.o.o. Zagreb, Croatia), with a micro piezo print head and aqueous pigment-based printing inks. The maximum print resolution of the machine is 1440 dpi. Machine properties and specifications are shown in Table 1. The printing inks were supplied under the commercial name Azon Pigment Ink, by Azonoprinter d.o.o. The composition of the printing ink is given in Table 2.

Table 1. Inkjet textile printing machine technical characteristics.

Print Head	Nozzle (Colour)	Nozzle (White)	Ink Cartridge	Max. Ink Drop Size	Print Speed
Piezo	192/per Colour	192 × 4	200 mL (with encrypted chip)	Up to 37 pL	Fast-28 s (C)/55 s (C + W) Normal-53 s (C)/102 s (C + W) Quality-79 s (C)/165 s (C + W)

Table 2. Printing inks composition.

Colour	Water (%)	Aliphatic Alcohol (%)	Ethylene Glycol (%)	Polyglycol Ether (%)	Polymers (%)	Pigment (%)	Triethylene Glycol Monobutyl Ether
Yellow	55–95	1–10	1–10	1–10	1–10	1–5	
Magenta	50–94	1–10	1–10	1–10	1–10	1–5	1–5
Cyan	55–95	1–10	1–10	1–10	1–10	1–5	
Black	55–95	1–10	1–10	1–10	1–10	1–5	

Both types of cotton fabrics (standard bleached and cationized) were printed without and with the addition of a surface acrylate-based binder–a polymer micro-emulsion 'Pigment Pre-treatment Solution' specially developed for digital printing by Azonprinter d.o.o. (Zagreb, Croatia). The composition of the binder used is shown in Table 3.

Table 3. Composition of binder used.

Components:	Water	Inorganic Nitrate	Acrylic Polymer	Formaldehyde
Ratio (%):	65–85	10–20	5–15	<0.02

Samples were prepared in sizes of 20 cm × 30 cm, and consumption of the polymer micro-emulsion (binder) over defined surfaces were 30 mL.

Samples were printed in two primary colours, magenta and cyan. By software adjustment of the prepress, a print with two different amounts of pigment was defined: in full concentration to achieve maximum saturation, which would correspond to the software definition of 100% and in concentration to achieve a shade of lower saturation, and increased lightness, which would correspond to the software definition of 50% of a pigment.

2.3. Microscopic Imaging

The microscopic imaging of printed samples was performed using a DinoLite AM7013 microscope. The imaging was performed before and after the 10th cycle of washing process.

2.4. Colourfastness

Printed samples were tested to washing fastness, in a laboratory apparatus for wet and dyeing processes Polycolor, Mathis. The test was performed according to standard ISO 105-C06:2010 (A2S) Textiles—Tests for colour fastness—Part C06: Colour fastness to domestic and commercial laundering, using 1 g/L of $Na_2BO_3H_2O$ and 4 g/L of standard detergent (James Heal ECE A, free from optical brighteners and phosphates), with a bath ratio of 1:200, temperature of 30 ± 2 °C, time of 40 min and pH of 6.

The results of colour fastness have been obtained objectively by spectrophotometric measurement using a DataColor Spectra Flash 600 PLUS–CT remission spectrophotometer (with constant instrument aperture, D65, using d/8° geometry) and were graded according to grey scale. The results are presented in terms of total colour difference (dE_{CMC}), according to the CMC(l:c) equation, accepted by ISO 105-J03:2009, for colour fastness and colour changes evaluation in the field of textiles [31]. The results were obtained after 1st, 3rd, 5th, 7th and 10th cycles of washing.

3. Results and Discussion

The purpose of cationic pre-treatment of cotton is to improve printability with pigment by introducing positively charged sites on cotton. Due to the pigment in the ink drop being ionized, the ink drops are negatively charged. Therefore, the ink drops could be immobilized on the cationized fabric surface. The print quality in terms of penetration, colour yield and colour depth have been analysed in comparison to untreated cotton and cotton pre-treated with conventional acrylate binder.

The aim of such research is to find an alternative to conventional binder application in order to achieve optimal colour properties, colour yield and colour fastness but with more satisfactory tactile and physical-mechanical fabric properties.

In order to compare standard and cationized cotton fabric from the aspect of surface charge and later analysis of the influence on the formation of surface pigment layer, the measurement of electro kinetic (zeta, ζ, ZP) potential was performed. In Figure 1, the zeta potential vs. pH of electrolyte 0.001 mol/L KCl is given.

Standard (unmodified) cotton fabric (B) has a negative zeta potential due to negative surface groups, hydroxyl of cellulose (-OH) and carboxyl that revels in chemical bleaching (-COOH). Therefore, ZP in whole pH range is negative, −24.7 mV at pH 9, and −10.0 mV at pH 3. No isoelectric point is present. Cationization during mercerization with Rewin DWR changes zeta potential. Ammonium groups (-NH$_2$) from polyammonium compound are present as well, resulting in higher zeta potential of −14.5 mV at pH 9. A change of zeta potential towards more positive values is visible, and such cationized cotton has an isoelectric point at 4.2. Cationized cotton has a positive charge at pH lower than 4.2.

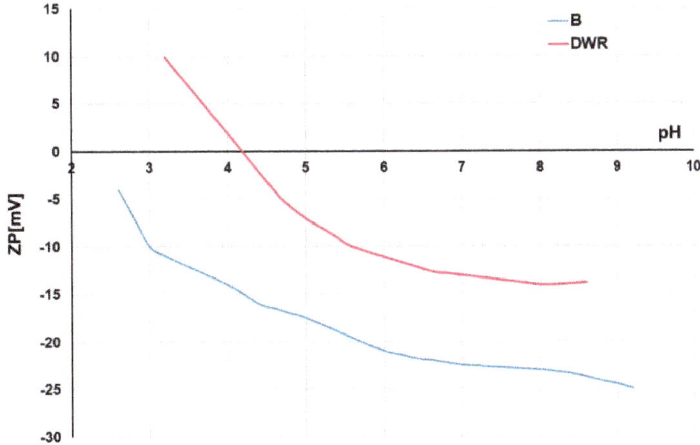

Figure 1. Zeta potential (ZP) vs. 0.001 mol/L KCl of cotton fabrics: standard unmodified (B) and cationized with Rewin DWR (DWR).

The mechanisms of penetration and spread of droplets are fundamental mechanisms of ink and textile surface relationships in digital printing processes. Therefore, the MMT and contact angle measurement on DSA30S were performed. The results are presented in Tables 4–7 and Figures 2 and 3.

Table 4. The results of moisture management testing for standard (non-cationized) cotton fabric presented as mean value.

		S0	S50	S100	SB0	SB50	SB100
AR (%/s)	T	44.03	45.00	71.67	89.13	41.07	112.09
	B	55.30	71.00	69.33	47.40	54.32	20.25
OMMC		0.50	0.43	0.37	0.32	0.37	0.04

Symbols: T-upper surface; B-lower surface; AR-absorption rate; OMMC-total ability to manage (liquid) moisture.

Table 5. Drop shape analyser (DSA) results for standard (non-cationized) cotton fabric.

S0		S50		S100		SB0		SB50		SB100	
CA	T	CA	T	CA	T	CA	T	CA	T	CA	T
/	/	/	/	/	/	49.76	42.00	76.82	28.00	81.16	15.00

Symbols: CA (m)-contact angle mean value; T-droplet absorption/spill time in seconds (s).

Table 6. The results of moisture management testing for cationized cotton fabric presented as mean value.

		C0	C50	C100	CB0	CB50	CB100
AR (%/s)	T	119.76	23.59	35.04	107.71	58.83	118.61
	B	17.51	29.10	40.18	18.35	48.88	29.24
OMMC		0.24	0.37	0.32	0.02	0.29	0.18

Symbols: T-upper surface; B-lower surface; AR-absorption rate; OMMC-total ability to manage (liquid) moisture.

Table 7. DSA results for cationized cotton fabric.

C0		C50		C100		CB0		CB50		CB100	
CA	T	CA	T	CA	T	CA	T	CA	T	CA	T
75.99	43	95.01	16	86.84	9	79.32	68	86.90	43	92.74	31

Symbols: CA (m)-contact angle mean value; T-droplet absorption/spill time in seconds (s).

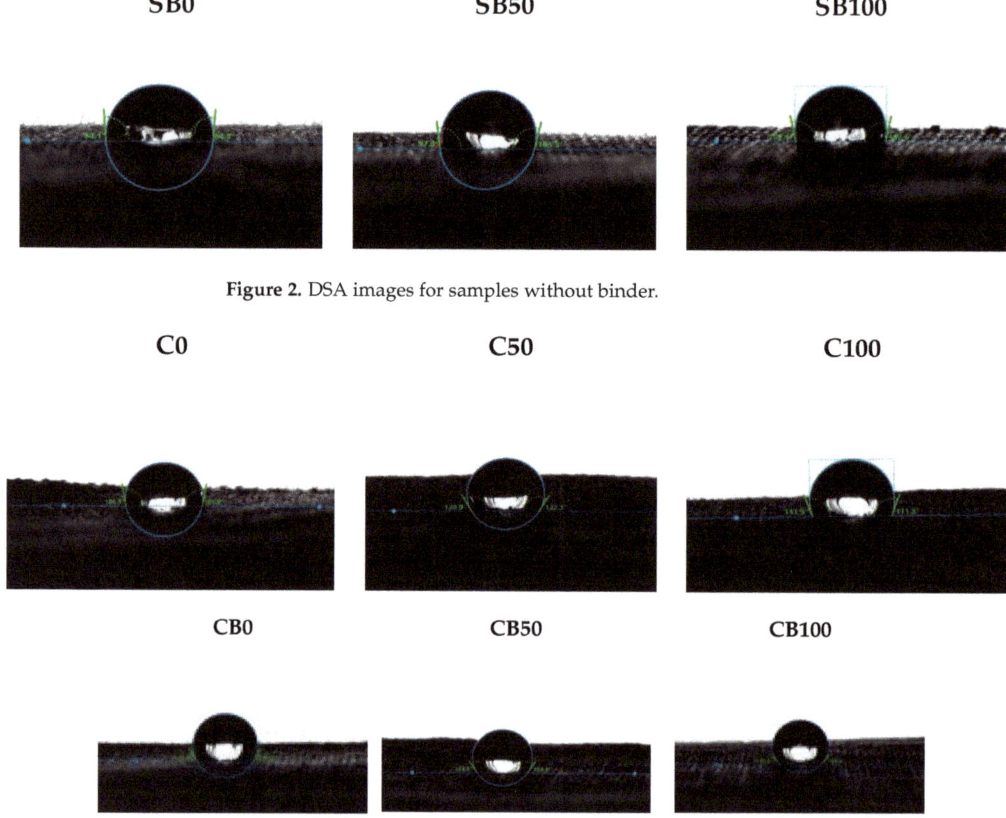

Figure 2. DSA images for samples without binder.

Figure 3. DSA images for samples cationized and cationized with binder.

For this part of the analysis, the samples are labelled as follows:

1. Non-printed, non-treated standard fabric-S0; standard fabric printed with 50% pigment-S50; standard fabric printed with 100% pigment-S100; standard fabric pre-treated with binder, non-printed-SB0; standard fabric pre-treated with binder, printed with 50% pigment-SB 50; standard fabric with binder, printed with 100% pigment-SB 100;

2. Non-printed, cationized fabric without binder-C0; cationized fabric without binder, printed with 50% pigment-C50; cationized fabric without binder, printed with 100% pigment-C100; cationized fabric with binder, non-printed-CB0; cationized fabric with binder, printed with 50% pigment-CB 50; cationized fabric with binder, printed with 100% pigment-CB 100.

Since the printing was performed with a commercial, closed settings system of inkjet printing inks, the moisture management testing as well as the contact angle measurement were performed with the aim of characterization of changes in substrate to be able to better explain the influence of textile pre-treatment on the formation, capillary spread and penetration of droplets into the structure of the material, as well as the passage of droplets through the material.

Moisture management testing is carried out to determine the absorbency, water repellence and water resistance of the fabric, which largely depends on the fabric structure itself, but also on the pre-treatment of the fabric. A comparative measurement of standard and cationized samples was performed, with and without the addition of a binder and with and without the presence of a pigment layer. In Table 4, the absorption rate (AR) and the total moisture management capability (OMMC) values are presented as indicators of moisture management. Considering evaluation as type of fabric, non-binder-treated samples (S0, S50, S100) are classified as fast absorbing and quick drying fabric, regardless of whether they are printed (regardless of the pigment presence in the surface structure of the fabric). In the printing process of such surfaces, due to the pronounced capillarity of the untreated textile, rapid absorption of the aqueous phase occurs. There is a deeper penetration of pigment into the substrate, and it does not remain on the surface but is unevenly incorporated into shallower and deeper layers of yarn structure and is not a factor in moisture management. Increased absorption capacity is observed in the S100 sample (sample printed with the maximum amount of pigment).

For samples pre-treated with binder, a change in the absorption rate is observed. The standard non-printed sample with binder SB0 and sample printed with 50% of pigment content SB50 belong to the group of textile substrates with the ability to penetrate moisture, while the sample SB100 belongs to the group of water-repellent substrates.

Drop shape analyser (DSA) measures the contact angle of the droplet on the surface of the fabric and the time of absorption/spillage of the droplet from the surface. The results of DSA show the significant influence of binder pre-treatment. As can be seen from Table 5, droplet analysis for standard samples without binder could not be performed, due to too fast passage of the test droplet through the fabric. This result correlates to MMT overall evaluation of fast absorbing and quick drying fabric. Only pre-treatment with a binder gives a slower absorbing surface that retains the droplet for a certain time. The contact angle of the droplet with the fabric surface is shown for samples pre-treated with binder (Figure 2).

In contrast to non-cationized samples, Table 6 immediately shows that all three cationized samples without the binder (C0, C50 and C100) belong to the group of textile substrates with the ability to manage moisture. It corresponds to the effect of simultaneous mercerization and cationization, whereby a larger number of active groups are formed in cotton and the sample can bind a larger amount of water. A significant value of the absorption rate was obtained for T (upper surface), which confirms the importance of cationization. The application of pigment reduces the number of polar molecules in cotton, responsible for binding water, and it is seen by the results obtained for the printed cationized samples (C50 and C100). Cationized samples pre-treated with binder (CB0, CB50, CB100), according to the results obtained (Table 6), are classified as water penetration to water repellent fabric, suggesting strong binding of acrylic binder. Results of DSA correspond to and confirm the MMT results. The obtained high contact angle indicates that the surface has low wetting properties.

In the next step, the samples were printed by digital inkjet technique of textile printing in two primary colours of the CMYK system, magenta and cyan, and the pigments were layered in concentrations of 50% and 100%. The appearance of samples is shown in Figure 4. Images of samples shown before washing confirm the positive effect of cationization of the cotton surface, on the coverage of the surface with pigment and the amount of bound pigment and the uniformity of the pigment layer. Even in the samples on which the conventional acrylate-based binder was applied, it is observed that the pre-treatment by cationization contributed to the achievement of greater coverage and more even distribution of the pigment layer. For the analysis of printed samples, unlike the previous analysis which included non-printed samples, new code abbreviations—labels are listed in Table 8 considering the pigment and pre-treatment.

In Figure 5, the value of the colour depth coefficient K/S for samples before washing is shown. The highest colour depth (K/S value) was obtained for samples treated with a binder, with the positive effect of cationization being obvious. However, the relevant indicators of the cationization effect can be seen by comparing the samples that were not treated with the binder, and for the cationized samples, a higher K/S is obtained, which indicates a higher amount of bounded pigment.

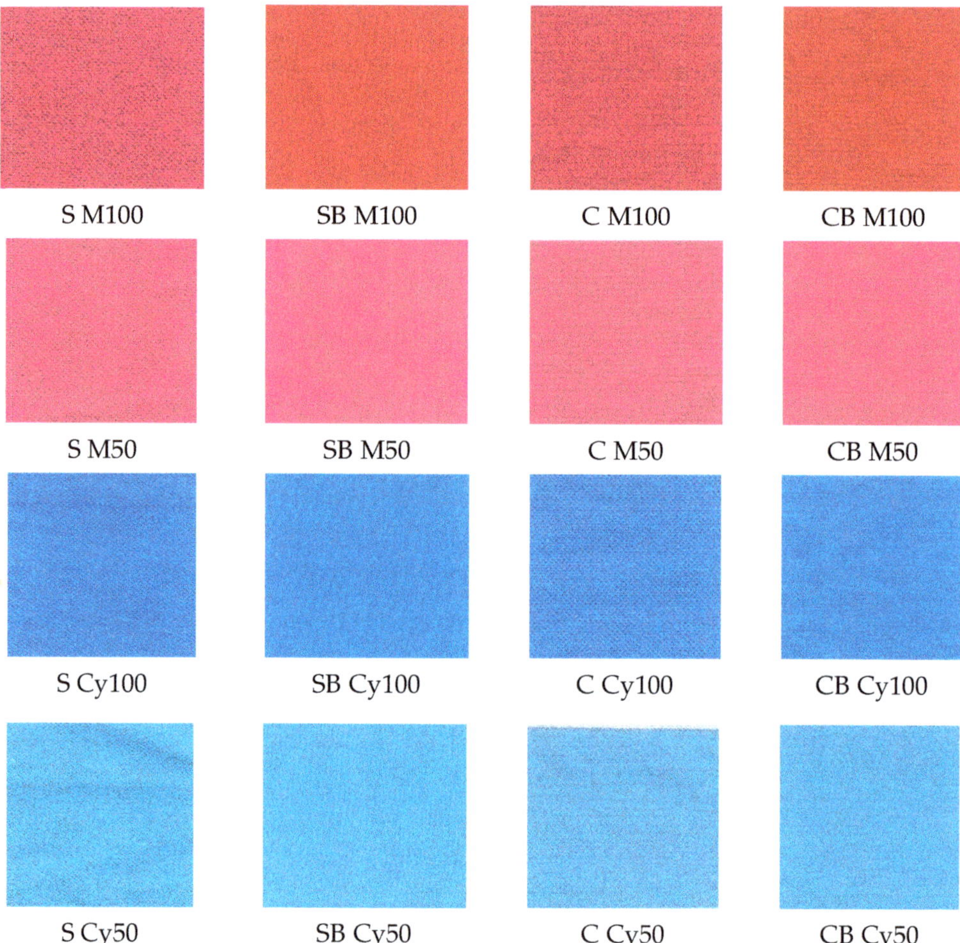

S M100	SB M100	C M100	CB M100
S M50	SB M50	C M50	CB M50
S Cy100	SB Cy100	C Cy100	CB Cy100
S Cy50	SB Cy50	C Cy50	CB Cy50

Figure 4. The appearance of printed samples.

Table 8. Sample descriptions and labels with code abbreviation.

Sample Descriptions	Label	
	Magenta	Cyan
Standard fabric, no binder, full pigment concentration (100%)	S M100	S Cy100
Standard fabric, no binder, full pigment concentration (50%)	S M50	S Cy50
Standard fabric, with binder, full pigment concentration (100%)	SB M100	SB Cy100
Standard fabric, with binder, full pigment concentration (50%)	SB M50	SB Cy50
Cationized fabric, no binder, full pigment concentration (100%)	C M100	C Cy100
Cationized fabric, no binder, full pigment concentration (50%)	C M50	C Cy50
Cationized fabric, with binder, full pigment concentration (100%)	CB M100	CB Cy100
Cationized fabric, with binder, full pigment concentration (50%)	CB M50	CB Cy50

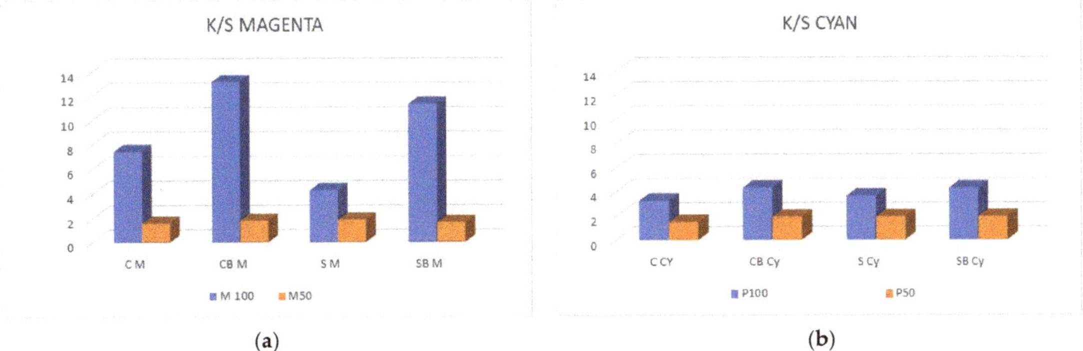

Figure 5. The K/S values of printed samples before washing. (**a**) Magenta and (**b**) Cyan pigment.

Tables 9 and 10 show the objective values of lightness (L*), chroma (C*) and hue (h*) of the measured samples. For the purpose of comparing the values of the colour parameters, the tables comparatively show the results for the non-cationized (labelled S) and cationized (labelled C) samples. In general, it can be seen from the results that the specific ratio of lightness and chroma, from which the colour intensity arises, depends on the interaction of the amount of pigment applied and the characteristics (pre-treatment) of the fabric.

Table 9. Objective values of colour parameters lightness (L*), chroma (C*) and hue (h°) for samples printed with magenta pigment.

		L*		C*		h°	
S M100	C M100	45.29	50.51	50.08	44.55	352.75	349.6
S M50	C M50	61.53	60.24	31.53	34.51	349.11	348.92
SB M100	CB M100	40.25	41.36	54.78	54.21	358.78	359.12
SB M50	CB M50	60.68	61.79	34.20	33.52	349.07	349.11

Table 10. Objective values of colour parameters lightness (L*), chroma (C*) and hue (h°) for samples printed with cyan pigment.

		L*		C*		h°	
SCy100	C Cy100	55.54	54.89	35.9	37.74	248.71	248.22
S Cy50	C Cy50	64.66	63.26	26.62	30.97	234.48	233.29
SB Cy100	CB Cy100	53.07	53.89	37.63	39.06	245.77	245.93
SB Cy50	CB Cy50	62.06	62.95	29.11	31.05	234.30	234.57

The dependence of the results on the specific nature of the colour itself is also visible. For samples printed in magenta pigment (Table 5), the lightness (L*) of a standard sample without a binder printed with 100% pigment is lower, and the chroma (C*) is higher compared to the cationized fabric. This can be explained by a more uniform coverage of the cationized surface with pigment. There is less scattering of incident light and during the measurement the objective value of lightness (L*) is higher and the chroma (C*) is lower. Also, one should take into account the nature of the magenta colour itself, which is a colour of medium lightness and takes on maximum chroma at medium levels of lightness. For samples with added binder, the lightness (L*) and chroma (C*) results for standard and cationized fabric, for both pigment concentrations, were uniform.

In the case of samples printed with pigment in cyan hue, a more pronounced influence of cationization is observed, and higher values of chroma (C*) are obtained in all cationized fabric, regardless of the pre-treatment with binder.

In the further work, the test of resistance to washing was performed, which was carried out in standard washing conditions in 10 cycles. Due to the extensiveness of the analysis and the amount of results, the microscopic images of the samples before and after the 10^t wash cycle will be presented, while the analysis in colour change in terms of total colour difference are shown for all washing cycles performed. The microscopic images for samples printed in magenta hue are shown in Figure 6, while the total colour differences (dE_{CMC}) are shown graphically in Figure 7. Also, the images of samples printed in cyan hue are shown on Figure 8 and the total colour differences for cyan-printed samples are shown in Figure 9.

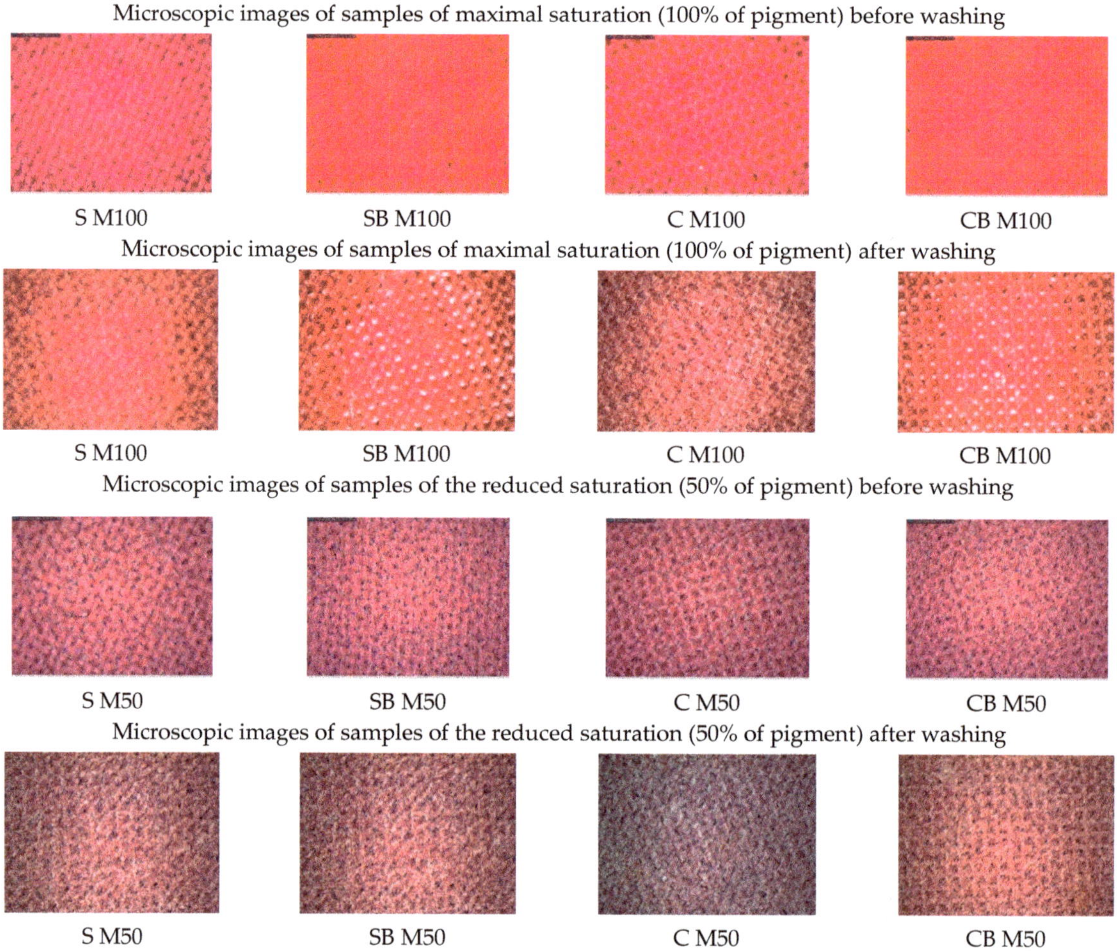

Microscopic images of samples of maximal saturation (100% of pigment) before washing

| S M100 | SB M100 | C M100 | CB M100 |

Microscopic images of samples of maximal saturation (100% of pigment) after washing

| S M100 | SB M100 | C M100 | CB M100 |

Microscopic images of samples of the reduced saturation (50% of pigment) before washing

| S M50 | SB M50 | C M50 | CB M50 |

Microscopic images of samples of the reduced saturation (50% of pigment) after washing

| S M50 | SB M50 | C M50 | CB M50 |

Figure 6. Microscopic images of samples printed in magenta colour hue, before and after the 10th washing cycle.

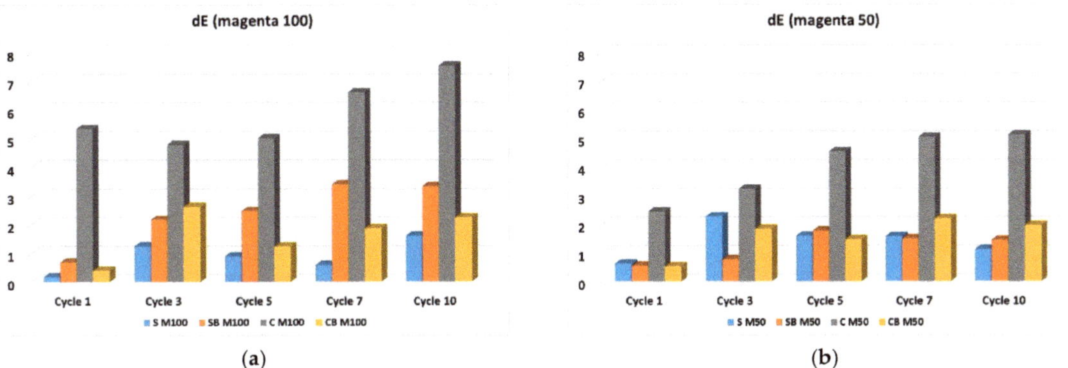

(a)

(b)

Figure 7. Total colour differences (dE) for washed samples compared with samples before washing, for magenta colour hue: (**a**) 100% (**b**) 50% of pigment.

Microscopic images of samples of maximal saturation (100% of pigment) before washing

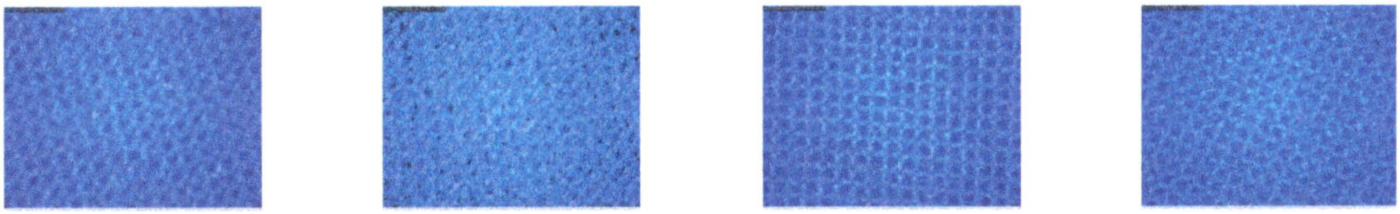

| S Cy100 | SB CY100 | C CY100 | CB CY100 |

Microscopic images of samples of maximal saturation (100% of pigment) after washing

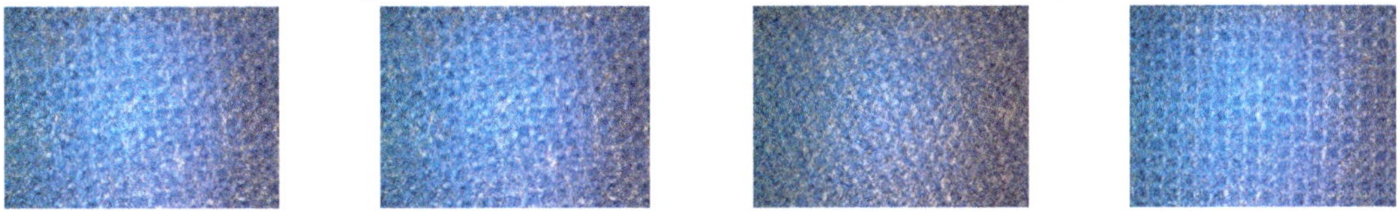

| S CY100 | SB CY100 | C CY100 | CB CY100 |

Microscopic images of samples of the reduced saturation (50% of pigment) before washing

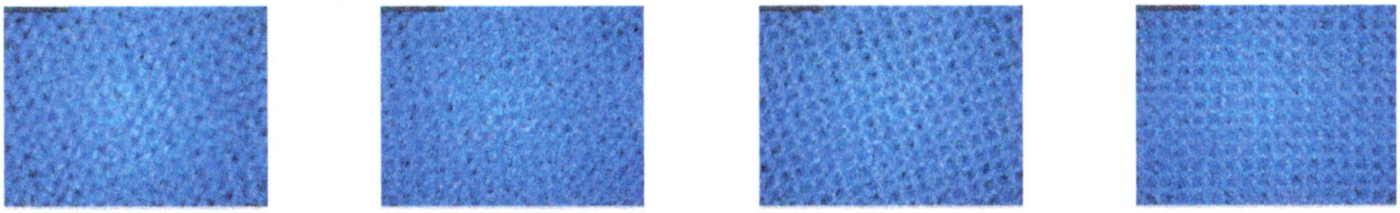

| S Cy50 | SB CY50 | C CY50 | CB CY50 |

Microscopic images of samples of the reduced saturation (50% of pigment) after washing

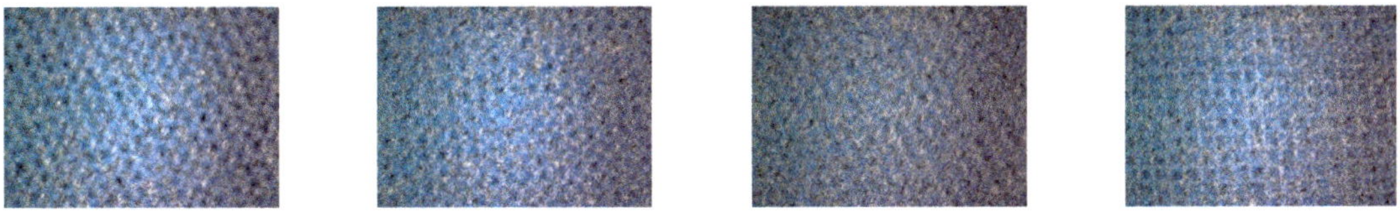

| S CY50 | SB CY50 | C CY50 | CB CY50 |

Figure 8. Microscopic images of samples printed in cyan colour hue, before and after the 10th washing cycle.

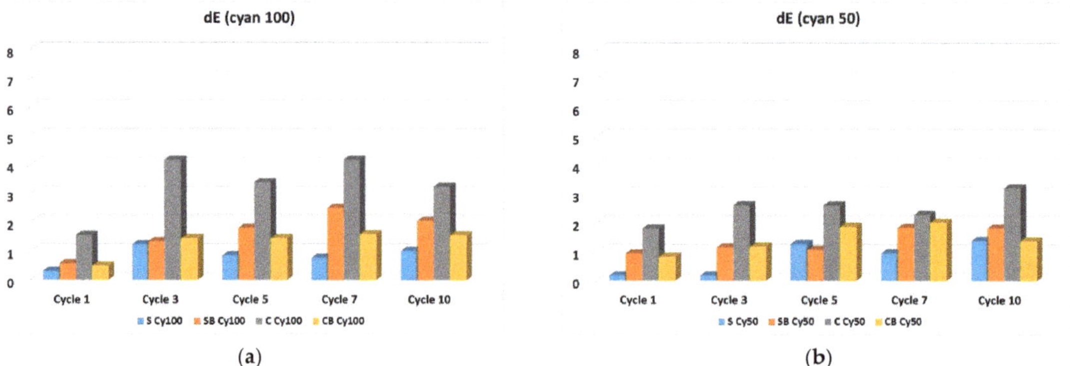

Figure 9. Total colour differences (dE) for washed samples compared with samples before washing, for cyan colour hue: (**a**) 100% (**b**) 50% of pigment.

Comparative analysis of microscopic images of samples printed with 50% and 100% pigment content confirmed that the cationized cotton achieves a more uniform pigment layer, higher colour efficiency with less pigment penetration into the textile structure, which is the goal to keep the pigment on the surface while preserving natural physical and mechanical properties of textile fabric.

Also, visual analysis of the washing effect reveals, in general, lower damage of the pigment layer in cationized fabrics. Namely, in the case of binder-free samples, there is a deeper penetration of pigment within the fabric structure. The influence of the surface structure of textiles is more pronounced, resulting in darker and less saturated colours. By applying the binder, a cross-linked three-dimensional polymer structure is created on the surface of the fabric, which keeps the pigment on the surface, and the entire layer is thicker, more even and the influence of the structure is less pronounced. However, due to the retention of the pigment layer on the textile surface, in samples with the addition of binders, both for standard and cationized fabric, there is more damage in washing cycles because there is cracking and peeling of the polymer layer. It is in these samples, with binder, that less damage and greater stability of the pigment layer is observed in cationized fabrics. This is because conventional binders of the older generation are of anionic character, and cationization of the fabric promotes the creation of a stronger bond between the polymer layer of binder and pigment with the fabric.

A comparison of the total differences (dE) as an analysis of the changes that occur in the pigment layer during the wash cycles shows the relative unevenness of the results. By comparing samples without binder, higher colour differences are obtained for cationized fabrics. The cationization of cotton led to the deposition of more pigment, and as expected, there was a significant release in the washing process, especially since there is no additive binder. However, in the case of samples with binder, printed with a higher amount of pigment (100%), lower values of the difference are obtained for cationized fabrics compared to non-cationized ones. For samples printed with a lower pigment concentration (50%), the results are uneven, and a detailed comparison cannot be performed. This is partly due to the fact that magenta pigment is a mixture of red and blue pigment. With lower amounts of pigment, i.e., with lighter colours, uneven removal of pigment occurs during washing, and there are significant changes in colour, which can be seen in microscopic images. Therefore, high values of the total colour difference will be the result of significant shifts in colour tone, and such differences cannot be attributed solely to lower fastness, but to certain changes in the spectral characteristics of colour.

In samples printed with pigment in cyan hue (Figure 8), the same effect of more uniform pigment layer and more even surface coverage in cationized fabrics is observed, in general, regardless of the amount of pigment and pre-treatment with binder. Also, less damage of the pigment layer due to washing is observed. Although samples printed with 100% pigment (more saturated colour of the print) show a significant change in colour after the 10th wash cycle, and obviously a certain amount of pigment is removed, the damage shown on microscopic images in the form of peeled parts of the polymer layer is less in cationized samples.

Also, in cyan, the lower wash fastness of the pigment layer is confirmed in cationized fabrics without binder compared to the non-cationized ones, but in samples with added binder, generally better pigment layer stability and wash fastness was obtained for cationized samples (Figure 9).

4. Conclusions

In the initial printing phase, the K/S values confirmed the obtained higher values of colour strength of cationized fabrics. The reason for the colour strength increase is that, as the cationic reagent concentration increases, the ionic attraction of the anionic pigment by the fibre's cationic charges also increases. The cationization results in better penetrating properties because of the reaction between the hydroxyl group of cotton and the cationic reagent. The wash colourfastness phase did not fully confirm the positive effect of cationization on the bonding of the pigment to the cotton fabric, but in the samples with binder, the positive effect of cationization on the stability and bond strength between the polymer layers as a pigment carrier with the cotton fabric was confirmed. This research is part of the extensive work that continues in the field of testing innovative methods of cotton processing in digital pigment printing processes.

Author Contributions: I.Č., T.D. and A.T. performed cationization; M.I.G. and P.L. performed printing; T.D. and P.L. performed washing and microscopic analysis; I.Č., A.T. and P.L. performed MMT and DSA; M.I.G. and P.L. performed colorimetric analysis; M.I.G., A.T. and T.D. designed the study; M.I.G. writing—original draft preparation. All authors have read and agreed to the published version of the manuscript.

Funding: This work has been supported in part by Croatian Science Foundation under the project UIP-2017-05-8780 HPROTEX and in part by University of Zagreb (TP4/21, TP6/21 and TP19/21).

Institutional Review Board Statement: Not applicable.

Informed Consent Statement: Not applicable.

Data Availability Statement: Data available in a publicly accessible repository.

Acknowledgments: A paper recommended by the 14th Scientific and Professional Symposium Textile Science and Economy, The University of Zagreb Faculty of Textile Technology.

Conflicts of Interest: The authors declare no conflict of interest.

Sample Availability: Samples of the compounds are available from the authors.

References

1. Colaris-Pigment Inkjet Printing for All Fibers. Available online: https://www.zimmer-austria.com/en/content/download-kufstein (accessed on 2 March 2021).
2. Ding, Y.; Zhendong, W.; Chuanxiong, Z.; Ruobai, X.; Wenliang, X. A Study of the Application of Pigment Printing on Cotton Fabrics. *Text. Res. J.* **2021**, *91*, 2283–2293. [CrossRef]
3. Ujiie, H. *Digital Printing of Textile*; Woodhead Publishing Limited & Textile Institute: Cambridge, UK, 2006.
4. Tawiah, B.; Howard, E.K.; Asinyo, B.K. The Chemistry of InkJet Inks for Digital Textile Printing—Review. *BEST IJIMITE* **2016**, *4*, 61–78.
5. Xue, H.C.; Shi, M.M.; Chen, H.Z.; Wu, G.; Wang, M. Preparation and Application of Nanoscale Microemulsion as Binder for Fabric InkJet Printing. *Colloids Surf. A Physiochem. Eng. Asp.* **2006**, *287*, 147–152. [CrossRef]
6. Kangwansupamonkon, W.; Suknithipol, M.; Phattanarudee, S.; Kiatkamjornwong, S. Inkjet Printing: Effects of Binder Particle Size and Chitosan Pretreatment on the Qualities of Silk Fabric. *Surf. Coat. Int.* **2011**, *94*, 216–225.
7. Momin, N. Chitosan and Improved Pigment Ink Jet Printing on Textiles. Available online: https://researchrepository.rmit.edu.au/esploro/outputs/doctoral/Chitosan-and-improved-pigment-ink-jet/9921861450501341 (accessed on 25 April 2020).

8. Ding, Y.; Shamey, R.; Chapman Parilla, L.; Freeman, H.S. Pretreatment Effects on Pigment-based Textile Inkjet Printing—Colour gamut and Crockfastness Properties. *Color. Technol.* **2019**, *135*, 77–86. [CrossRef]

9. Wang, L.; Hu, C.; Yan, K. A one-step Inkjet Printing Technology with Reactive Dye Ink and Cationic Compound Ink for Cotton Fabrics. *Carbohydr. Polym.* **2018**, *197*, 490–496. [CrossRef]

10. Li, M.; Zhang, L.; An, Y.; Ma, W.; Fu, S. Relationship between Silk Fabric Pretreatment, Droplet Spreading, and Ink-jet Printing Accuracy of Reactive Dye Inks. *J. Appl. Polym. Sci.* **2018**, *135*, 46703–46714. [CrossRef]

11. Kaimouz, A.W.; Wardman, R.H.; Christie, R.M. The Inkjet Printing Process for Lyocell and Cotton Fibres. Part 1: The Significance of Pre-treatment Chemicals and Their Relationship with Colour Strength, Absorbed Dye Fixation and Ink Penetration. *Dyes Pigm.* **2010**, *84*, 79–87. [CrossRef]

12. Yang, H.; Fang, K.; Liu, X.; Cai, Y.; An, F. Effect of Cotton Cationization Using Copolymer Nanospheres on Ink-Jet Printing of Different Fabrics. *Polymers* **2018**, *10*, 1219. [CrossRef]

13. Rekaby, M.; Abd-El Thalouth, J.I.; Abd El-Salam, S.H. Improving Reactive Printing via Cationization of cellulosic linen fabric. *Carbohydr. Polym.* **2013**, *98*, 1371–1376. [CrossRef]

14. Tabba, A.H.; Hauser, P.J. Effect of a Cationic Pretreatment on Pigment of Cotton Fabric. *TCC ADR* **2000**, *32*, 30–33.

15. Kanik, M.; Hauser, P.J. Printing of Cationized Cotton with Reactive Dyes. *Color. Technol.* **2002**, *118*, 300–306. [CrossRef]

16. Hauser, P.J. Printing of Cationized Cotton with Acid Dyes. *AATCC Rev.* **2003**, *3*, 25–28.

17. Kanik, M.; Hauser, P.J. Ink-Jet Printing of Cationised Cotton Using Reactive Inks. *Color. Technol.* **2003**, *119*, 230–234. [CrossRef]

18. Kanik, M.; Hauser, P.J. Printing of Cationized Cotton with Direct Dyes. *Text. Res. J.* **2004**, *74*, 43–50. [CrossRef]

19. Wang, C.; Zhao, S.; Wang, X. Improving the Color Yield of Ink-jet Printing on Cationized Cotton. *Text. Res. J.* **2004**, *74*, 68–71.

20. Wang, C.X.; Zhang, Y.H. Effect of Cationic Pretreatment on Modified Pigment Printing of Cotton. *Mater. Res. Innov.* **2007**, *11*, 27–30. [CrossRef]

21. Wang, C.; Yin, Y.; Wang, X.; Bu, G. Improving the Color Yield of Ultra-fine Pigment Printing on Cotton Fabric. *AATCC Rev.* **2008**, *8*, 41–45.

22. El-Shishtway, R.M.; Nassar, S.H. Cationic pretreatment of Cotton Fabric for Anionic dye and Pigment Printing with Better Fastness Properties. *Color Technol.* **2002**, *118*, 115–120. [CrossRef]

23. Grancarić, A.M.; Tarbuk, A.; Dekanić, T. Elektropozitivan pamuk. *Tekstil* **2004**, *53*, 47–51.

24. Tarbuk, A.; Grancarić, A.M.; Leskovac, M. Novel cotton cellulose by cationisation during mercerisation—Part 2: Interface phenomena. *Cellulose* **2014**, *21*, 2089–2099. [CrossRef]

25. Sutlović, A.; Glogar, M.I.; Čorak, I.; Tarbuk, A. Trichromatic Vat Dyeing of Cationized Cotton. *Materials* **2021**, *14*, 5731. [CrossRef] [PubMed]

26. Tarbuk, A.; Grancarić, A.M. Chapter 6 in Cellulose and Cellulose Derivatives: Synthesis, Modification and Applications, Part I: Cellulose Synthesis and Modification. In *Interface Phenomena of Cotton Cationized in Mercerization*; Nova Science Publishers: New York, NY, USA, 2015; pp. 103–126.

27. Čorak, I.; Brlek, I.; Sutlović, A.; Tarbuk, A. Natural Dyeing of Modified Cotton Fabric with Cochineal Dye. *Molecules* **2022**, *27*, 1100. [CrossRef] [PubMed]

28. Grancarić, A.M.; Tarbuk, A.; Jančijev, I. Dyeing Effects of Cationized Cotton. In *Color: Ciencia, Artes, Proyecto y Enseñanza*; Caivano, J.L., Lopez, M., Eds.; Grupo Argentino del Color: Buenos Aires, Argentina, 2006; pp. 39–44.

29. Tarbuk, A.; Sutlović, A.; Dekanić, T.; Grancarić, A.M.; Zdjelarević, I. Ultra-Deep Black Cationized Cotton by Metal-Complex Dyeing. In Proceedings of the 16th World Textile Conference AUTEX, Ljubljana, Slovenia, 8–10 June 2016; Simončič, B., Tomšič, B., Gorjanc, M., Eds.; CD-ROM 5–3. University of Ljubljana, Faculty of Sciences and Engineering: Ljubljana, Slovenia, 2016; pp. 1–7.

30. Grancarić, A.M.; Tarbuk, A.; Pušić, T. Electrokinetic Properties of Textile Fabrics. *Color Technol.* **2005**, *121*, 221–227. [CrossRef]

31. *ISO 105-J03*; Textiles-Tests for colour fastness-Part J03. Calculation of Colour Differences. International Organization for Standardization: Geneva, Switzerland, 2009.

Article

Silver Nanowires and Silanes in Hybrid Functionalization of Aramid Fabrics

Alicja Nejman [1,2], Anna Baranowska-Korczyc [1], Katarzyna Ranoszek-Soliwoda [2], Izabela Jasińska [1], Grzegorz Celichowski [2] and Małgorzata Cieślak [1,*]

[1] Łukasiewicz Research Network–Textile Research Institute, Brzezinska St. 5/15, 92-103 Lodz, Poland; alicja.nejman@iw.lukasiewicz.gov.pl (A.N.); anna.baranowska-korczyc@iw.lukasiewicz.gov.pl (A.B.-K.); izabela.jasinska@iw.lukasiewicz.gov.pl (I.J.)

[2] Department of Materials Technology and Chemistry, Faculty of Chemistry, University of Lodz, Pomorska St. 163, 90-236 Lodz, Poland; katarzyna.soliwoda@chemia.uni.lodz.pl (K.R.-S.); grzegorz.celichowski@chemia.uni.lodz.pl (G.C.)

[*] Correspondence: malgorzata.cieslak@iw.lukasiewicz.gov.pl

Abstract: New functionalization methods of *meta-* and *para*-aramid fabrics with silver nanowires (AgNWs) and two silanes (3-aminopropyltriethoxysilane (APTES)) and diethoxydimethylsilane (DEDMS) were developed: a one-step method (mixture) with AgNWs dispersed in the silane mixture and a two-step method (layer-by-layer) in which the silanes mixture was applied to the previously deposited AgNWs layer. The fabrics were pre-treated in a low-pressure air radio frequency (RF) plasma and subsequently coated with polydopamine. The modified fabrics acquired hydrophobic properties (contact angle Θ_W of 112–125°). The surface free energy for both modified fabrics was approximately 29 mJ/m^2, while for reference, *meta-* and *para*-aramid fabrics have a free energy of 53 mJ/m^2 and 40 mJ/m^2, respectively. The electrical surface resistance (R_s) was on the order of 10^2 Ω and 10^4 Ω for the two-step and one-step method, respectively. The electrical volume resistance (R_v) for both modified fabrics was on the order of 10^2 Ω. After UV irradiation, the Rs did not change for the two-step method, and for the one-step method, it increased to the order of 10^{10} Ω. The specific strength values were higher by 71% and 63% for the *meta*-aramid fabric and by 102% and 110% for the *para*-aramid fabric for the two-step and one-step method, respectively, compared to the unmodified fabrics after UV radiation.

Keywords: *para*-aramid; *meta*-aramid; silver nanowires; plasma surface modification; polydopamine; surface free energy; fabric durability; UV protection; hydrophobicity; conductive fabric

Citation: Nejman, A.; Baranowska-Korczyc, A.; Ranoszek-Soliwoda, K.; Jasińska, I.; Celichowski, G.; Cieślak, M. Silver Nanowires and Silanes in Hybrid Functionalization of Aramid Fabrics. *Molecules* **2022**, *27*, 1952. https://doi.org/10.3390/molecules27061952

Academic Editors: Ana Sutlović, Sanja Ercegović Ražić and Marija Gorjanc

Received: 22 February 2022
Accepted: 14 March 2022
Published: 17 March 2022

Publisher's Note: MDPI stays neutral with regard to jurisdictional claims in published maps and institutional affiliations.

1. Introduction

Due to their low density, high tensile strength, excellent thermal stability, high specific strength, and modulus of elasticity, aramid fibers are widely used in special clothes accessories and technical products [1]. Their weakness, however, is their poor resistance to UV radiation. Aramids absorb UV radiation in the range of 300 to 400 nm, which results in breaking of the intermolecular bonds in the polymer that causes deterioration of the mechanical properties [2,3]. Therefore, it is important to protect them against UV radiation such as by modification with nanosilver, which also provides antibacterial and antistatic properties. Such functionalization significantly extends the scope of their application, such as materials for thermal sensors in smart clothes, for the protection of electronic communication, and in aviation [4–6]. The molecular structure of aramid fibers, which consist of aromatic rings and amide groups, makes them highly crystalline [7,8]. Due to the smooth and chemically inert surface, they show poor adhesion to modifiers. In order to improve the functionalization efficiency, an initial surface modification is performed through plasma activation and deposition of a polydopamine thin film by oxidative polymerization of dopamine [9]. The activation increases the free surface energy of aramid fibers and creates more reactive sites, enabling better binding between the polydopamine and the aramid surface [9–11]. Polydopamine is a biopolymer that forms a coating on the fiber surface through induced dopamine oxidative self-polymerization. This process enables the creation of carboxyl, amine, and catechol functional groups in order to strengthen the bond between the fiber and the modifier and improve the physical properties of the composites [12–14]. Polydopamine shows adhesion for most types of inorganic and organic surfaces, including precious metals, oxides, polymers, semiconductors, and ceramics [13,15]. Studies on the adhesion mechanism of the polydopamine film show that functional groups create covalent and non-covalent connections with inorganic or organic materials [15–18]. This type of preliminary modification increases the deposition efficiency of silver nanoparticles and improves the durability of the coating [9,19–21]. Silver nanowires (AgNWs) are one type of nanomodifier used for the multifunctionalization of textile materials. They have two UV absorption peaks; the peak at 350 nm can be attributed to surface plasmon resonance (SPR) along the silver nanowires, and the peak at about 380 nm can be considered the transverse direction of the SPR [22,23]. AgNWs absorb UV radiation, constituting a protective barrier for the fibrous structure covered with them, but they can also degrade. Air humidity has a significant influence on the degradation under the influence of UV radiation [24]. For higher air humidity, the oxygen and water contained in the air under the influence of UV radiation can create ozone and hydroxyl groups, which can initiate oxidation, hydrolysis, and photodegradation processes of the AgNWs [24,25]. For protection against UV radiation, silanes are used [26–29] because they have the ability to reflect UV radiation [28]. Hydrophobic silane-based textile materials are usually produced by surface treatment with mixtures of silane agents and silane cross-linkers [28,30]. One of the silanes used for the modification is 3-aminopropyltriethoxysilane (APTES), which contains a polar and chemically active aminopropyl group and three easily hydrolysable alkoxy groups. The hybrid organo-silica film obtained by APTES hydrolysis is hydrophilic, and the water contact angle measured by Zeng et al. [31] for the APTES film was 50°. The addition of diethoxydimethylsilane (DEDMS) to tetraethoxysilane (TEOS) before the hydrolysis increased the hydrophobicity of the obtained coatings was studied by Zhang et al. [32]. Souza et al. [33], for a mixture of APTES with (3-glycidoxypropyl) trimethoxysilane (GPTMS) (ratio of 2:1), obtained a coating with a water contact angle of 109.7°. Hasanzadeh et al. [34] coated a polyester-viscose fabric with silica nanoparticles prepared from TEOS to obtain a rough and hydrophobic surface. The fabric was then treated with functionalized polydimethylsiloxane (PDMS) to reduce the surface free energy. This fabric became hydrophobic with a contact angle of 151°. APTES is most often used as a stabilizer and a fastener to immobilize silver nanoparticles on the fiber surface [35].

In this work, *meta-* and *para*-aramid fabrics were modified with AgNWs and a mixture of silanes, APTES, and DEDMS. There were two methods of modification tested, a one-step method using a mixture of silanes with a AgNWs colloid (ratio of 1:10) and a two-step method (layer-by-layer), in which a mixture of silanes was applied on the AgNWs coating. The aim of the research was to protect aramid fabrics against UV radiation, improve the durability, conductivity, and hydrophobic properties, and assess which modification method is more effective.

Obtained results will have a significant impact on the development of knowledge in the field of material engineering, surface physico-chemistry, and functionalization of aramid materials. The functionalization of aramid textile materials with AgNWs and silanes is an original strategy of improving functionality of aramid fabrics, by giving them mechanical resistance to UV radiation, hydrophobicity, and conductivity. Protection against UV radiation extends the usage of modified aramid fabrics, which has a beneficial effect on the environment. In addition, hydrophobicity reduces the contamination, including biofouling, and protects AgNWs against corrosion. In turn, imparting conductive properties aims to create conductive paths between elements and to protect electronic components against electrostatic discharge (ESD) in textronic products. Thanks to the combination of the functional properties of the AgNWs, silanes, and aramids, we developed modified aramid fabrics, which can be used in many fields, e.g., in the automotive and military industry, in protective clothing in high-risk occupations, to monitor vital functions or the level of available oxygen, and also in everyday life as textile materials for tents, umbrellas, and covering fabrics, etc.

2. Materials and Methods

2.1. Materials

Meta-aramid (poly (isophthalates-1,3-fenylodiamid)) (*m*Ar) fabric with a plain-weave and mass per unit area of 205 g/m^2 made of 25 × 2 tex yarn (number of threads: in warp—230/10 cm, in weft—160/10 cm), (Newstar®, Yantai Tayho Advanced Materials Co., Ltd., Yantai, China) and *para*-aramid (poly (1,4-terephthalate-fenylodiamid)) (*p*Ar) fabric with a plain-weave and mass per unit area of 165 g/m^2 made of 20 × 2 tex yarn (number of threads: in warp—240/10 cm, in weft—150/10 cm), (Kevlar®, DuPont, Londonderry, UK) were studied. Both fabrics were made in Łukasiewicz Research Network-Textile Research Institute using the harness loom (CCI Tech. Inc., New Taipei City, Taiwan).

2.2. Methods

2.2.1. Functionalization Methods

Preparation of Silver Nanowires (AgNWs) Colloid

The synthesis of AgNWs was described in a previous work [36]. In brief, AgNWs were synthesized by reduction of silver nitrate ($AgNO_3$, 99.9%, Sigma Aldrich, Gillingham, UK) with ethylene glycol ($C_2H_6O_2$, Avantor, Gliwice, Poland) in the presence of polyvinylpyrrolidone (PVP, 55,000 g/mol, Sigma Aldrich, UK).

The synthesis was performed in a Radleys 1l Reactor-Ready (Radleys, Essex, United Kingdom) reactor equipped with a mechanical stirrer, working at a speed of 150 RPM. The reactor has a double heating jacket, allowing for precise temperature control performed by an external thermostat during the entire synthesis. The temperature was set at $170 \pm 2\ °C$.

A total of 600 mL of ethylene glycol, 30 g of PVP, and 0.42 g of sodium chloride (NaCl, CHEMPUR, Piekary Śląskie, Poland) were heated to 167 °C (constant temperature control) and stirred at 150 RPM to obtain a homogenous solution. Then, a mixture of 6.12 g of $AgNO_3$ in 300 mL of ethylene glycol was added with a peristaltic pump at a rate of 3.33 mL/min for 90 min into the heated solution of ethylene glycol, PVP, and NaCl. When the silver precursor solution was added, the colloid was stirred at 150 RPM and heated for 60 min at 167 °C.

After the synthesis, the solution was air-cooled to about 25 °C. To remove ethylene glycol and excess PVP from the colloid, the subsequent, small parts of the solution were diluted with acetone (C_3H_6O, 99,5%, Avantor, Gliwice, Poland) with a ratio of 1:10 and shaken for 10 min. Then, the silver nanowires were dispersed in anhydrous ethyl alcohol (C_2H_6O, 99,8%, Avantor, Gliwice, Poland). The obtained AgNWs colloid concentration was 4000 ppm, the length of the AgNWs was about $10 \pm 2\ \mu m$, and the diameter was about 42 ± 3 nm.

Low-Pressure Air RF Plasma Treatment of Aramid Fabrics

Para-aramid and *meta*-aramid fabrics were washed in diethyl ether for 30 min and dried at 25 °C. Next, both fabrics were treated with a low-pressure air RF plasma (Zepto plasma system, Model 2. Diener Electronic, Ebhausen, Germany) for 10 min. The pressure in the plasma chamber was set at 0.3 mbar with laboratory air (Linde Gaz, Kraków, Poland) with the purity: 20% of O_2 and 80% of N_2 (40% relative humidity) as the working gas. The power of the plasma was set at 50 W and the generator frequencies was 40 kHz.

Polydopamine Functionalization of Aramid Fabrics

A dopamine solution with a concentration of 2 g/L was prepared by dissolving dopamine hydrochloride (Sigma Aldrich, UK) in distilled water. The pH of the solution was adjusted to 8.3 by adding tris/glicyne buffer (BIO RAD, Warszawa, Poland). The plasma treated aramid fabrics were immersed in the dopamine solution for 24 h (Figure 1a). Under the same conditions, a polydopamine coating was prepared on the surface of the aluminum plate. This material was used for the acquisition of a reference FTIR spectrum of pure polydopamine. After that, all samples were rinsed with distilled water three times and then dried for 24 h at 25 °C.

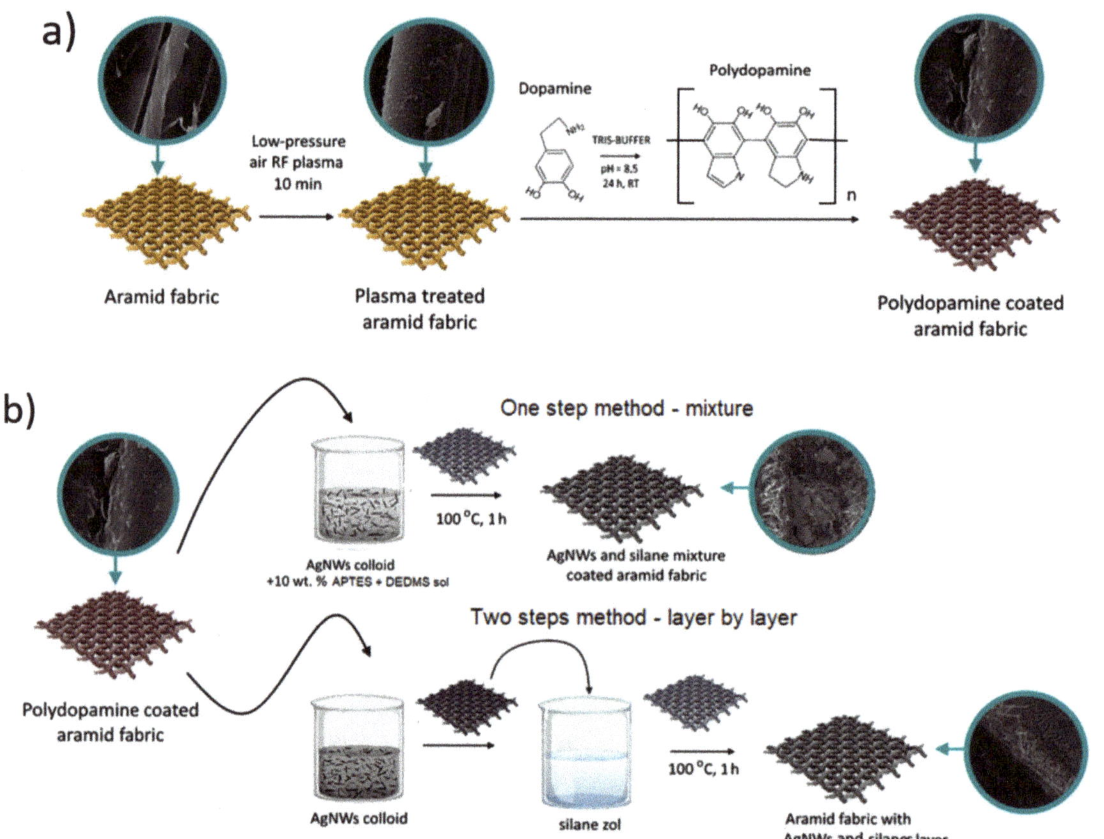

Figure 1. The scheme of the modification of aramid fabrics with (**a**) low-pressure air RF plasma and polydopamine and (**b**) the application of AgNWs and silanes by the mixture and layer-by-layer method.

Silanes Sol Preparation

APTES (Unisil, Tarnów, Poland) and DEDMS (Sigma Aldrich, Gillingham, United Kingdom) differ in their chemical structure, brittleness, and hydrophobicity. To prepare the AgNWs and silane modified aramid fabrics, silanes sol was prepared by mixing 15.00 g of APTES, 1.12 g of DEDMS, and 3.93 g of 1 M HCl (Avantor, Gliwice, Poland). The silanes mixture was shaken for 30 min. Next, ethyl alcohol (C_2H_6O, 99.8%, Avantor, Gliwice, Poland) was added at 50% by volume of the silanes mixture.

AgNWs and Silanes Functionalization of Aramid Fabrics

The one-step (mixture) and two-step (layer-by-layer) methods were used for the functionalization of aramid fabrics with AgNWs and silanes (Figure 1b). In the one-step method, aramid fabrics were 5-fold dipped in a mixture of a AgNWs colloid (Ag) with 10 wt.% silanes sol (S) for 1 min. After the mixture was applied every time, the fabrics were dried in an oven at 100 °C for 1 h. The fabrics modified with the mixture (Ag + S) are denoted as mAr/RF/PD/5Ag+S and pAr/RF/PD/5Ag+S for the *meta-* and *para*-aramid fabric, respectively.

In the two-step method, fabrics were dipped in the AgNWs colloid for 1 min and dried at 25 °C. The AgNWs colloid (Ag) was applied 5 times. Then, the fabrics were dipped 1 time in the silanes sol (S) and then dried in an oven at 100 °C for 1 h. The fabrics modified with the layer-by-layer (Ag/S) method are denoted as mAr/RF/PD/5Ag/S and pAr/RF/PD/5Ag/S for the *meta-* and *para*-aramid fabric, respectively.

Moreover, as control samples, aramid fabrics were modified with AgNWs. A 1-fold (mAr/RF/PD/1Ag and pAr/RF/PD/1Ag) and 5-fold (mAr/RF/PD/5Ag and pAr/RF/PD/5Ag) colloid application was used.

In Table 1, the values of mass per unit area of the reference and functionalized aramid fabrics are summarized.

Table 1. Mass per unit area of the reference and modified aramid fabrics.

Fabric	Mass per Unit Area, g/m^2	Mass per Unit Area of AgNW Coating, g/m^2	Mass per Unit Area of Silanes Coating, g/m^2	Mass per Unit Area of AgNWs and Silanes Coating, g/m^2
mAr	205 ± 2			
mAr/RF/PD/1Ag	214 ± 6	9 ± 3		
mAr/RF/PD/5Ag	248 ± 7	43 ± 2		
mAr/RF/PD/5Ag+S	273 ± 7			68 ± 5
mAr/RF/PD/5Ag/S	305 ± 9	43 ± 2	62 ± 1	105 ± 3
pAr	165 ± 3			
pAr/RF/PD/1Ag	171 ± 4	6 ± 1		
pAr/RF/PD/5Ag	200 ± 1	35 ± 3		
pAr/RF/PD/5Ag+S	222 ± 5			57 ± 2
pAr/RF/PD/5Ag/S	263 ± 4	35 ± 3	63 ± 5	98 ± 6

UV Irradiation

The UV irradiated unmodified and modified fabrics were placed between two (100 W) LED diodes with the maximum wavelength of the emitted light at 365 nm for 96 h. The irradiance (W) was determined by using the CUV 4 irradiance sensor (Kipp & Zonen B.V., Delft, Netherlands) to measure the intensity of the light in the UV range (305–385 nm). The supplied irradiation energy (P) was calculated from Equation (1) [37] and determined to be 46,719 kJ/m^2 for one LED diode.

$$P = \frac{t \cdot 3.6 \cdot \Sigma W}{n} \tag{1}$$

where P is the irradiation energy (kJ/m^2), t is the irradiation time (h), W is the irradiance (W/m^2), and n is the number of measurements during the irradiation time t.

2.2.2. Characterization Methods
SEM/EDS Analysis

The morphology and chemical composition analysis were performed using a scanning electron microscope (SEM) (Nova NanoSEM 450 FEI, Hillsboro, OR, USA) and VEGA 3 (TESCAN, Brno, Czech Republic) with an energy dispersive spectroscopy (EDS) X-ray microanalyzer INCA Energy (Oxford Instruments Analytical, High Wycombe, United Kingdom) with a magnification of 2500×, 10,000×, and 20,000×. A total of three EDS spectra were recorded for each sample, and mean values of the weight percentage of the elements were determined.

Optical Microscopy

The images of the fabric surface before and after the abrasion process were acquired using an optical microscope DSX1000 (OLYMPUS, Tokyo, Japan) and using the DEPH function to increase the depth of field. A magnification of 300× was used for all images.

FTIR Analysis

FTIR/ATR (Fourier transform infrared spectroscopy/attenuated total reflectance) spectra of reference and plasma-treated yarns were recorded in the range 600–4000 cm^{-1} using Nicolet IS 50 spectrometer (Thermo Fisher Scientific Inc., Bartlesville, OK, USA) with GATR (grazing angle attenuated total reflectance) accessory (Harrick Scientific Products Inc., New York, NY, USA) using the MCT (mercury-cadmium-telluride) detector. This accessory allows to collecting spectra from the surface of the fibers at a depth of about up to 50 nm.

Surface Properties

The analysis of the surface properties of the fabrics was performed by the goniometric method using the goniometer PGX (Fibro System AB, Stockholm, Sweden). In order to determine the surface free energy (γ_s), three standard liquids with known surface tensions and different values of dispersive and polar components were applied (Table 2). The drop of liquid with a volume of 4 µL was applied. A total of three repetitions for each sample were used. The surface free energy was determined according to the Wu model (Equation (2)), which is based on the assumptions of the Owens-Wendt model [7,38–42] but describes the intermolecular interaction using the harmonic mean.

$$\gamma_l(1+\cos\theta) = 4((\gamma_l{}^d\gamma_s{}^d)/\gamma_l{}^d + \gamma_s{}^d + \gamma_l{}^P\gamma_s{}^P/\gamma_l{}^P + \gamma_s{}^P) \text{ and } \gamma_s = \gamma_s{}^d + \gamma_s{}^P \tag{2}$$

where $\gamma_s{}^d$ and $\gamma_s{}^P$ are the dispersive and polar components of a solid, respectively, and $\gamma_l{}^d$ and $\gamma_l{}^P$ are the dispersive and polar components of a liquid, respectively.

Table 2. Characteristics of standard liquids.

Liquid	Surface Tension, mJ/m^2		
	γ_l	$\gamma_l{}^d$	$\gamma_l{}^P$
Water (distilled)	72.8	21.8	51.0
Formamide (99.5%, Sigma-Aldrich, UK)	58.0	39.0	19.0
Diiodomethane (99%, Sigma-Aldrich, UK)	50.8	48.5	2.3

Specific Strength

The study of the specific strength of the fabrics before and after UV irradiation was performed by using the Instron 3367 Test Machine (United Kingdom) in accordance with PN-EN ISO 13934-1:2013-07 Textiles—Tensile properties of flat products—Part 1: Determination of maximum force and relative elongation at maximum force by the strip method. A total of five repetitions for each sample were used. Samples were conditioned for 24 h at a temperature of 21.0 ± 1.0 °C and a relative humidity (RH) of 64.4 ± 0.1% and then tested in the same conditions.

Abrasion Resistance

The resistance to abrasion of the fabrics before and after UV irradiation was studied using the Martindale unit (United Kingdom) in accordance with PN-EN ISO 12947-2:2017-02 Textiles—Determination of the abrasion resistance of fabrics by Martindale method—Part 2: Determination of specimen breakdown. A total of three repetitions for each sample were used. Samples were conditioned for 24 h at a temperature of 21.1 ± 0.1 °C and an RH of 64.4 ± 0.1% and were tested in the same conditions. A standard woolen fabric was used as an abrasive cloth. An abrasive load of 12 kPa was applied.

Conductivity

The electrical surface resistance (R_s) and electrical volume resistance (R_v) of the fabrics before and after UV irradiation were determined according to the PN-EN 1149-1:2008 for R_s and PN-EN 1149-2:1999+A1:2001 for R_v, using a set of standard electrodes and a 6206 teraohmmeter (ELTEX, Weil am Rhein, Germany). The fabrics were conditioned for 24 h in the HCZ 0030 L(M) chamber (Heraeus, Hanau, Germany) at a temperature of 23.0 ± 2.0 °C and an RH of 25.0 ± 5.0% and were tested in the same conditions.

3. Results and Discussion

3.1. Modification of the Aramid Fabrics with Low-Pressure Air RF Plasma and Polydopamine

The surface of the *meta-* (Figure 2a) and *para*-aramid (Figure 2a) fibers is smooth with visible longitudinal cracks. Aramid fibers have a fibrillar structure. Activation in low-pressure RF plasma causes an increase in the unevenness and surface roughness (Figures 1b and 2b).

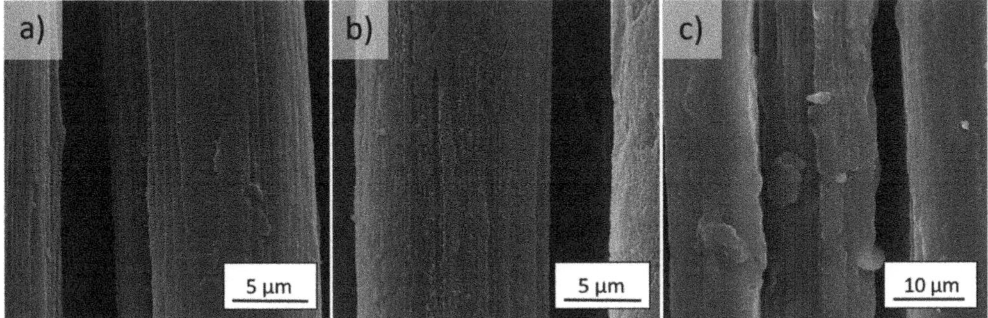

Figure 2. SEM images of (**a**) reference, (**b**) plasma treated, and (**c**) polydopamine modified *m*Ar fibers (magnification of 10,000×).

The dopamine solution prepared for the functionalization of the plasma-activated aramid fabrics was colorless and transparent. During the oxidative autopolymerization of dopamine, the color of the solution quickly turned pink as catechol was oxidized to benzoquinone. Then, the pink solution slowly turned to dark brown. This indicates that after polymerization, the reaction of melanin formation takes place, resulting in the creation of polydopamine (PD) with a high strength of irreversible covalent bonds on the matrix. After modification, polydopamine formed a layer on the surface of both types of fibers (Figures 2c and 3c).

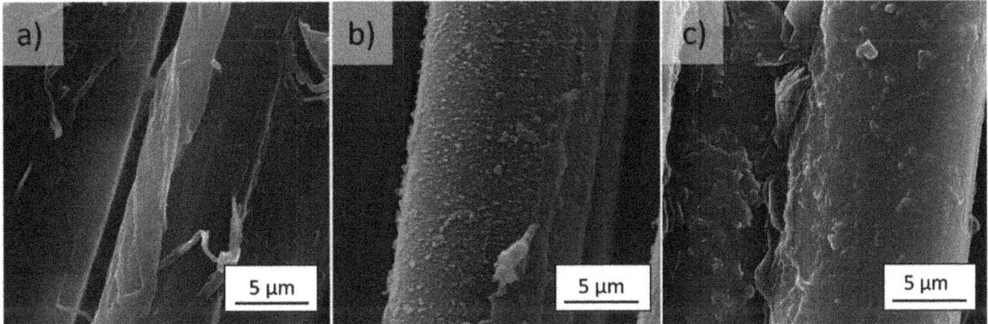

Figure 3. SEM images of (**a**) reference, (**b**) plasma treated, and (**c**) polydopamine modified *p*Ar fibers (magnification of 10,000×).

The presence of the polydopamine coating on the aramid fiber surface was observed during the SEM analysis and confirmed by the FTIR spectroscopy results (Figure 4). In Table 3, the characteristic bands and their wavenumbers for the *p*Ar fabric and polydopamine are presented. After application of the dopamine coating on the *p*Ar fabric, an increase in the band intensity at 698 cm^{-1}, corresponding to the C–H bond out-of-plane of the substituted aromatic ring, and at 1513 cm^{-1}, corresponding to the C=N bond stretching vibrations of the aromatic ring, in relation to the unmodified *p*Ar (Figure 3a) is observed. In the spectrum for the *p*Ar/RF/PD fabric, bands are present in the spectrum for polydopamine but not in the spectrum for the reference *p*Ar fabric. These are the bands at 911 cm^{-1} characteristic of the bending vibrations of the CONH bond and CN stretching, at 1440 cm^{-1} corresponding to the stretching vibrations of the C=C bond of the aromatic ring (Figure 4a), and at 3175 cm^{-1} corresponding to the stretching vibrations of the N–H bond (Figure 4b).

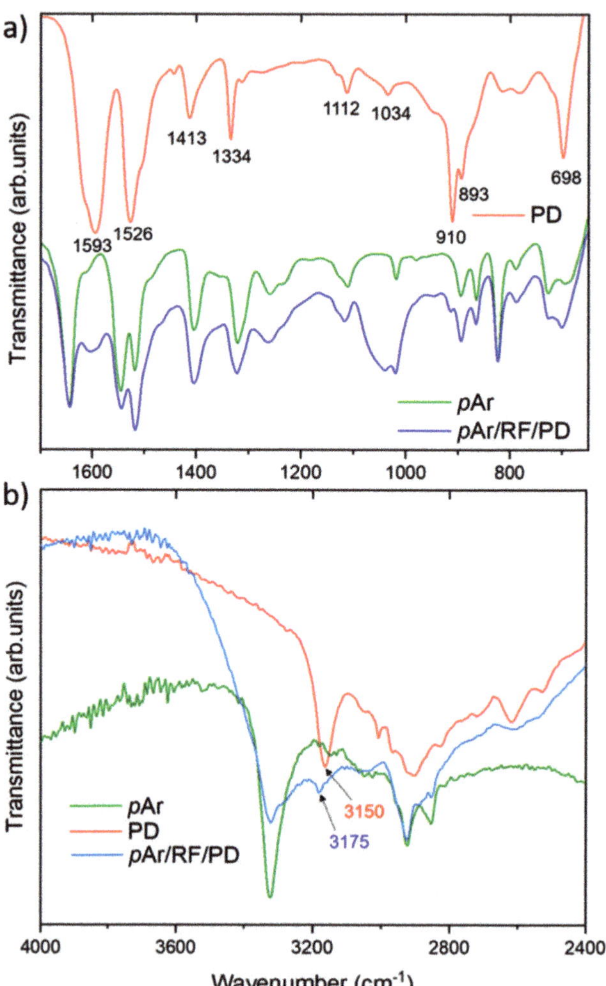

Figure 4. FTIR spectrum of polydopamine, reference, and RF plasma and polydopamine functionalized *para*-aramid fabrics in the range of (**a**) 600–1750 cm^{-1} and (**b**) 2400–4000 cm^{-1}.

Table 3. FTIR bands assigned for the pAr fabric and the reference polydopamine film.

*p*Ar Fabric		Reference Polydopamine Film	
Bands, cm^{-1}	Description [7,10,34]	Bands, cm^{-1}	Description [7,10,34]
698	C–H out of plane substituted aromatic ring	698	C–H out of plane substituted aromatic ring
821	*p*-substituted phenyl	911	CONH bending, C–N stretching
1306	C–N stretching	1034	C–O stretching
1509	C=C stretching	1113	C–H bending
1538	N–H bending	1335	C–N stretching
1637	C=O stretching	1413	C=O stretching
1740	C=O stretching	1440	C=C stretching vibrations of an aromatic ring
3323	N–H stretching	1527	C=N stretching vibrations of an aromatic ring
		1594	C=C stretching vibrations of an aromatic ring, C–N stretching
		3150	N–H stretching

3.2. AgNWs and Silanes Modified Aramid Fabrics before and after UV Irradiation

SEM analysis shows that the surface of both aramid fibers after the 1-fold application of AgNWs (*m*Ar/RF/PD/1Ag and *p*Ar/RF/PD/1Ag) is covered unevenly (Figures 5a and 6a). After the 5-fold application, the coating is uniform (Figure 4b), with few areas without AgNWs (Figure 6b). AgNWs are found both on the surface of the fibers and inside the fabric structure. After application of the Ag+S mixture (Figures 5c and 6c), the AgNWs were embedded inside the silanes coating and deposited both on the surface of the fibers and in the spaces between them. In the case of the layer-by-layer method (Figures 5d and 6d), a layer of silanes covers the surface of the fabric with AgNWs. There are also fibers with AgNWs that protrude above the silanes surface.

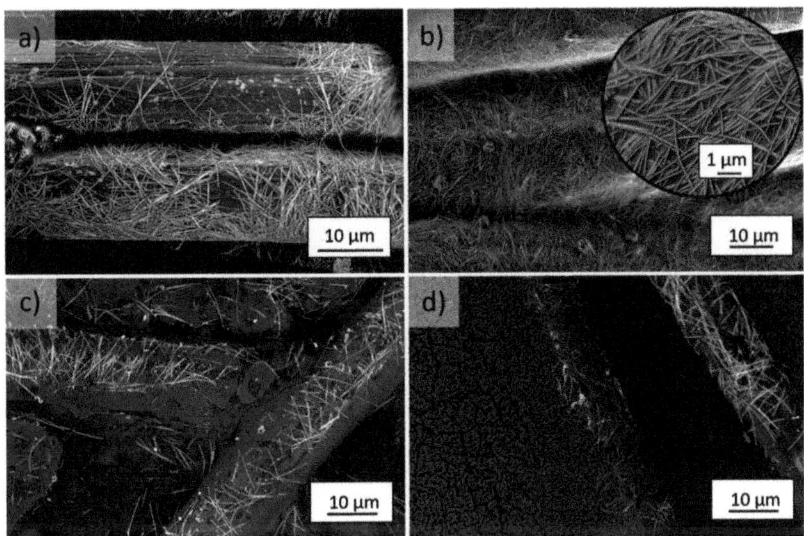

Figure 5. SEM images of modified *m*Ar fibers after the (**a**) 1-fold and (**b**) 5-fold application of AgNWs and the (**c**) mixture Ag+S, and (**d**) layer-by-layer Ag/S modification (magnification of 2500× and 20,000×).

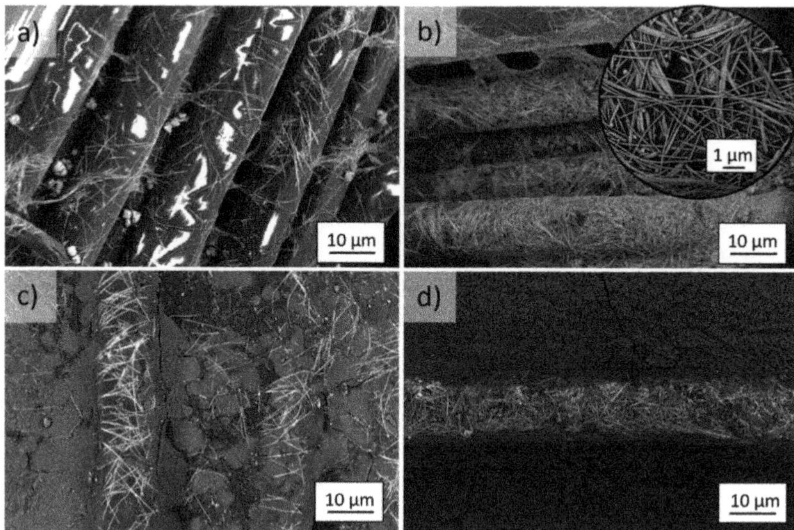

Figure 6. SEM images of modified *p*Ar fibers after the (**a**) 1-fold and (**b**) 5-fold application of AgNWs and the (**c**) mixture Ag+S, and (**d**) layer-by-layer Ag/S modification (magnification of 2500× and 20,000×).

EDX analysis of the *m*Ar and *p*Ar fabric shows the presence of C, N, and O (Table 4). For the *m*Ar fabric, the content of C, N, and O is 68, 11, and 20 wt.%, respectively. For the *m*Ar/RF/PD/1Ag fabric, the content of Ag is 3 wt.%, and for *m*Ar/RF/PD/5Ag, it is 6-fold higher (Table 4). With the increase in the multiplicity of the AgNWs application, the content of C, N, and O decreased, which proves the effective coverage of the aramid fibers with AgNWs. For *m*Ar/RF/PD/5Ag+S, the C and N content does not change significantly, and the O content, which is derived from the ethoxyl groups in silanes, increases by about 50% compared to *m*Ar/RF/PD/5Ag. A decrease in the C content by 26% and an increase in the O and N content by 70% and 24%, respectively, are noted for *m*Ar/RF/PD/5Ag/S in relation to *m*Ar/RF/PD/5Ag. These changes are due to the presence of the silanes layer, which contains ethoxy groups and an amino group. The Ag content on the fiber surface is about 60% and 30% lower for *m*Ar/RF/PD/5Ag+S and *m*Ar/RF/PD/5Ag/S compared with *m*Ar/RF/PD/5Ag. The Si content is 5 wt.% and 7 wt.% for *m*Ar/RF/PD/5Ag+S and *m*Ar/RF/PD/5Ag/S, respectively.

The content of C, N, and O for the *p*Ar fabric is 66, 13, and 19 wt.%, respectively. In the case of *p*Ar/RF/PD/1Ag, the Ag content is 6 wt.%, and after the 5-fold application, it is 3.5 times higher. The content of C, N, and O decreases with the increase in the number of AgNWs applications, which indicates the effective coverage of the fibers with AgNWs. For *p*Ar/RF/PD/5Ag+S, the C and N content stays at the same level in relation to *p*Ar/RF/PD/5Ag. The O content increases by 48%, which is due to the presence of ethoxyl groups in the silanes. For *p*Ar/RF/PD/5Ag/S, the C content is lower by 22%, and the N and O content is higher by 46% and 72%, respectively, which, similarly to the *meta*-aramid fabric, is due to the surface being covered with silanes. For *p*Ar/RF/PD/5Ag+S and *p*Ar/RF/PD/5Ag/S, the Ag content on the fiber surface is lower by about 64% and 42%, respectively, than for *p*Ar/RF/PD/5Ag. The Si content is 5 wt.% for *p*A/RF/PD/5Ag+S and 6 wt.% for *p*Ar/RF/PD/5Ag/S.

Table 4. The mean values of the weight percentages of the elements with standard deviations for unmodified and modified *m*Ar and *p*Ar fabrics.

Fabric	Weight Percentage of Elements, wt.%				
	C	N	O	Si	Ag
*m*Ar [1]	68 ± 1	10 ± 1	21 ± 1		
*m*Ar/RF/PD/1Ag [2]	68 ± 1	8 ± 1	20 ± 1		3 ± 1
*m*Ar/RF/PD/5Ag [2]	56 ± 2	7 ± 1	18 ± 1		18 ± 2
*m*Ar/RF/PD/5Ag+S [2]	54 ± 1	6 ± 1	28 ± 1	5 ± 1	6 ± 1
*m*Ar/RF/PD/5Ag/S [2]	41 ± 1	8 ± 1	31 ± 1	7 ± 1	12 ± 1
*p*Ar [1]	66 ± 1	13 ± 1	19 ± 1		
*p*Ar/RF/PD/1Ag [1,2]	66 ± 1	6 ± 1	20 ± 1		6 ± 1
*p*Ar/RF/PD/5Ag [1,2]	54 ± 3	5 ± 1	18 ± 1		21 ± 3
*p*Ar/RF/PD/5Ag+S [1,2]	54 ± 1	5 ± 1	26 ± 1	5 ± 1	8 ± 1
*p*Ar/RF/PD/5Ag/S [1,2]	42 ± 1	7 ± 1	31 ± 1	6 ± 1	12 ± 1

[1] The total weight percentage of the elements is less than 100 wt.% because of the trace amounts of Na and S (S only for the *p*Ar fabric), which are less than 1 wt.% and are not included in Table 2. In the case of *m*Ar, the presence of Na is due to the technology of fiber production as a result of the reaction of two comonomers in tetrahydrofuran. An oligomer suspension is formed, which combines with sodium carbonate, and fibers are formed [7,25]. In turn, *p*Ar fibers are produced by the condensation of 1,4-diaminobenzene and terephthaloyl chloride. The formed *p*Ar is immersed in sulfuric acid, and then sodium hydroxide is added for neutralization. The product of neutralization is sodium sulphate [7,25]. [2] For all AgNWs modified fabrics, the trace amounts of Cl (<1 wt.%), which came from silver chloride as a by-product of the AgNWs colloid synthesis, were present and are not included in Table 2.

After 96 h of UV irradiation of the unmodified *m*Ar (Figure 7a) and *p*Ar (Figure 8a) fabrics, the roughness of the fibers surface increased, and polymer fragments were present on the surface. The UV irradiation caused degradation of the AgNWs on the fabrics, as evidenced by silver precipitates on their surface (Figure 7b insert, Figure 8b insert). The coating on the *m*Ar and *p*Ar fibers modified with the Ag + S mixture was cracked (Figures 7c and 8c), and coating defects were visible, which proves that fragments of the Ag + S coating were detached (Figure 7c). UV radiation causes breaking of the polysiloxane bonds and photo-oxidation, which induces the formation of silanol and carbonyl groups [27]. In the case of the layer-by-layer method, the AgNWs did not degrade and did not change their morphology (Figures 7d and 8d). In the mixture method, the AgNWs were more exposed to the UV radiation due to the thin coating of silanes compared with the layer-by-layer method. This may cause the degradation of AgNWs and damages in silanes coating.

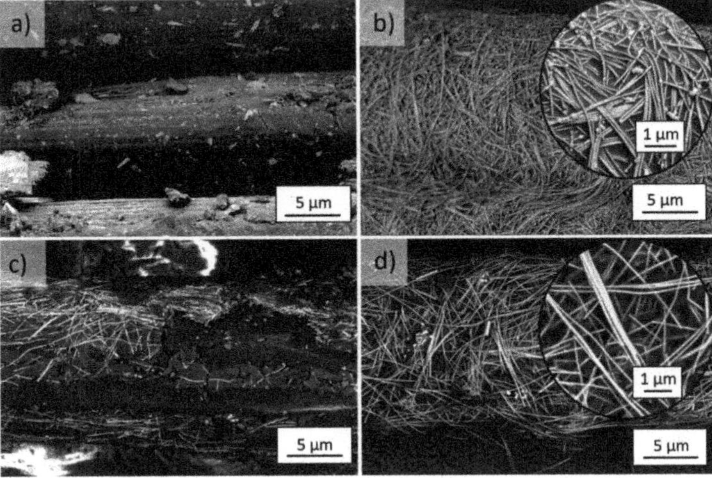

Figure 7. SEM images of UV irradiated (365 nm, 96 h) *m*Ar fibers (**a**) reference and after (**b**) 5-fold AgNWs application, (**c**) mixture Ag+S and (**d**) layer by layer Ag/S modification (magnification of 6500× and 20,000×).

Figure 8. SEM images of UV irradiated (365 nm, 96 h) *p*Ar fibers (**a**) reference and after (**b**) 5-fold AgNWs application, (**c**) mixture Ag+S, and (**d**) layer-by-layer Ag/S modification (magnification of 6500× and 20,000×).

3.3. Surface Properties

Table 5 presents the values of contact angles for water (Θ_W), formamide (Θ_F), and diiodomethane (Θ_{DIM}) for the unmodified and modified fabrics.

Table 5. Contact angles of water (Θ_W), formamide (Θ_F), and diiodomethane (Θ_{DIM}) for unmodified and modified *m*Ar and *p*Ar fabrics.

Fabric	Contact Angle [°]		
	Θ_W	Θ_F	Θ_{DIM}
*m*Ar	64 ± 2	58 ± 2	63 ± 2
*m*Ar/RF	19 ± 3	36 ± 3	27 ± 3
*m*Ar/RF/PD	0 ± 0	36 ± 1	90 ± 0
*m*Ar/RF/PD/1Ag	77 ± 2	74 ± 3	36 ± 4
*m*Ar/RF/PD/5Ag	87 ± 1	90 ± 2	32 ± 3
*m*Ar/RF/PD/5Ag+S	125 ± 5	106 ± 2	55 ± 2
*m*Ar/RF/PD/5Ag/S	112 ± 2	96 ± 2	51 ± 1
*p*Ar	77 ± 4	33 ± 2	14 ± 3
*p*Ar/RF	12 ± 5	18 ± 5	5 ± 1
*p*Ar/RF/PD	0 ± 0	19 ± 1	68 ± 1
*p*Ar/RF/PD/1Ag	84 ± 4	77 ± 2	36 ± 3
*p*Ar/RF/PD/5Ag	89 ± 3	85 ± 2	34 ± 4
*p*Ar/RF/PD/5Ag+S	120 ± 1	106 ± 3	58 ± 2
*p*Ar/RF/PD/5Ag/S	114 ± 2	100 ± 1	53 ± 1

For the *m*Ar fabric, the Θ_W value is 64°, and after plasma activation, it decreases by 70%. For the *p*Ar fabric, the Θ_W value is 20% higher than for the *m*Ar fabric. Plasma treatment reduced the Θ_W value by 84%. The decrease in the Θ_W value of *p*Ar/RF is greater than for *m*Ar/RF because of the chemical structure of both aramids and the higher degree of crystallinity of *para*-aramid, whose structure is more rigid and orderly, which causes greater changes on the fibers surface as a result of plasma etching [7]. After the application of the polydopamine coating, for the *m*Ar/RF/PD and *p*Ar/RF/PD fabrics, the value of Θ_W is 0°. The decrease in the Θ_W value is due to the presence of active polar functional groups on the surface, such as –OH, –NH$_2$, or –NH (Figure 4), which interact with the polar liquid. A decrease in the Θ_W value was also observed by Sabdin et al. [42], who modified polypropylene (PP) meshes by immersing them in 10 mM dopamine buffered with TRIS at pH 8.5 for 24 h. Then, the PP meshes were washed several times with distilled water and dried at 40 °C. The Θ_W value decreased from 138.9° to 74.1°. Jiang et al. [43] immersed a PP membrane for 24 h in a 1.0 g/L aqueous dopamine solution prepared by dissolving 0.25 g of dopamine in 250 mL of deionized water. Then, 1 mL of 1 M NaOH was added to the solution to adjust the pH to 9. The Θ_W value decreased from 117.5° to 54.3°. Chen et al. [44] observed a decrease in the Θ_W value from 143.7° to 134.2° after modification of the PP nonwoven fabric in an aqueous solution of dopamine hydrochloride at a concentration of 2 g/L and buffered with TRIS (pH = 8.5) for 24 h at 60 °C and then the samples were washed four times with ultrasounds and dried in a vacuum at 60 °C. This indicates the exposure of hydroxyl and amine groups in polydopamine on the nonwoven surface, which causes the increase in the wettability.

The Θ_W values for both aramid fabrics with AgNWs are comparable and higher than for the unmodified fabrics, and their hydrophobicity increases with the increasing number of AgNWs deposited.

Fabrics modified with the mixture and layer-by-layer methods are hydrophobic. The *m*Ar/RF/PD/5Ag+S and *p*Ar/RF/PD/5Ag+S fabrics have the highest Θ_W values, which are 125° and 120°, respectively. For the *m*Ar/RF/PD/5Ag/S and *p*Ar/RF/PD/5Ag/S fabrics, the Θ_W values are 112° and 114°, respectively. Zeng et al. [31] reported that the water contact angle of the APTES film is 52°, which means that APTES is hydrophilic. In turn, Zhang et al. [32] measured the contact angle of a DEDMS/TEOS silica aerogel (ratio of 0.26), which was 139.9°, while TEOS is hydrophilic [45].

The value of the surface free energy (γ_s) for the *m*Ar fabric is 39.8 mJ/m^2, which is lower than for the *p*Ar fabric by about 35% (Figure 9a,b). Plasma treatment and modification with polydopamine resulted in a significant increase in the γ_s values by about 68% and 84%, respectively, for *m*Ar and by about 87% and 83% for *p*Ar fabrics in relation to the unmodified fabrics. After the 1- and 5-fold AgNWs application, the γ_s value slightly decreased with the increase in the modification multiplicity for both types of aramid fabrics and is slightly lower in relation to the unmodified fabrics. The γ_s values for the aramid fabrics modified with silanes by both methods are a little lower than for the fabrics after the 5-fold AgNWs application.

The dispersion component value (γ_s^d) of the *m*Ar fabric is 23.8 mJ/m^2, which is lower than that of the *p*Ar fabric by about 50%. The plasma treatment causes a 52% increase and 21% decrease in the γ_s^d component for the *m*Ar/RF and *p*Ar/RF fabric, respectively. The decrease in γ_s^d for the *p*Ar fabric after plasma treatment in comparison with the *m*Ar fabric is related to the higher degree of crystallinity and the fibrillization phenomenon of the reference *p*Ar fabric (Figure 3a). For *m*Ar fabric surface, the enlargement of the surface roughness is observed after plasma treatment (Figure 2b), which causes the increase in the γ_s^d value. After modification with polydopamine, the value of the γ_s^d component decreases in relation to the unmodified fabrics as result of the formation of a layer of polydopamine on the surface of the fibers (Figures 2c and 3c). For both fabrics, after the 1-fold AgNWs application, the γ_s^d values are comparable. The lower γ_s^d value for *p*Ar/RF/PD/1Ag compared with the reference fabric is caused by the fibrillization process, which significantly affects the dispersion component value. For the 5-fold AgNWs modified fabrics, the γ_s^d values are still comparable and lower than for the 1-fold AgNWs modification, which may be related to the smoothing of the discontinuous surface after successive AgNWs applications. For both aramid fabrics modified with silanes by the mixture and layer-by-layer method, the γ_s^d values are lower in relation to the fabrics after the 5-fold AgNWs application. Slightly higher γ_s^d values are found for the layer-by-layer method, which may be because of the visible AgNWs coated fibers that protrude above the smooth silane surface.

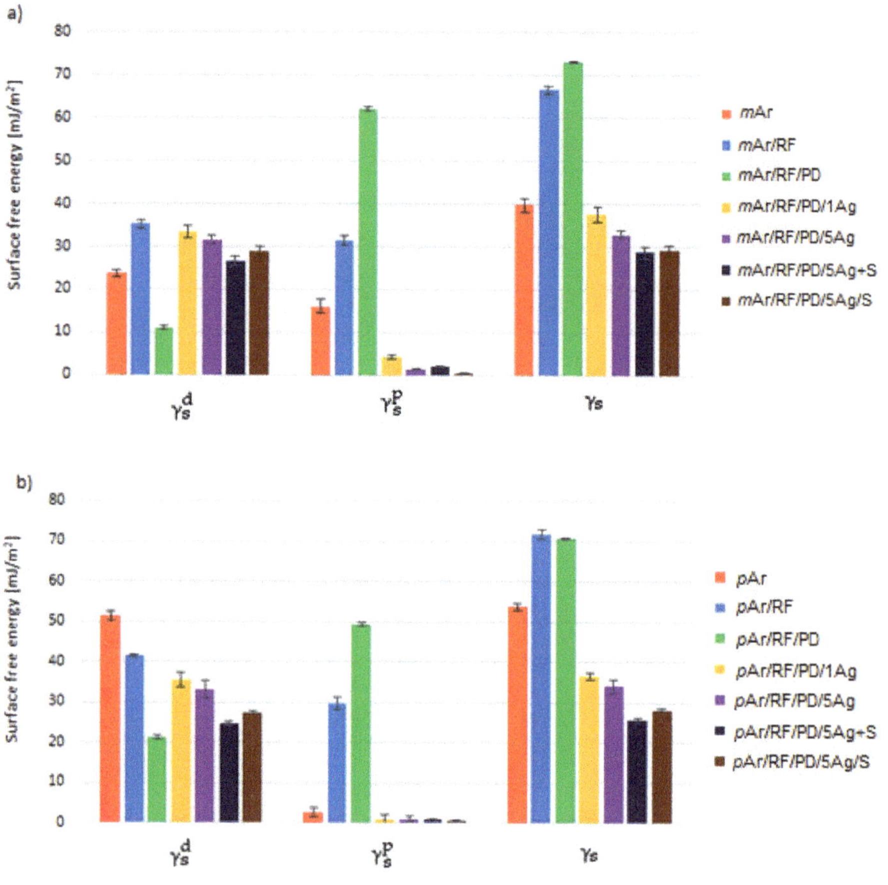

Figure 9. Surface free energy (γ_s) of unmodified and modified *m*Ar (**a**) and *p*Ar (**b**) fabrics (γ_s^{d}—dispersive component, γ_s^{P}—polar component).

The value of γ_s^{P} for *m*Ar is 16.0 mJ/m^2, which is six times higher than the value for *p*Ar. After plasma treatment, the γ_s^{P} values increase and are comparable for both fabrics. The polydopamine application causes a further increase in the γ_s^{P} values by a factor of 2 and 1.7 for *m*Ar/RF/PD and *p*Ar/RF/PD, respectively. For fabrics modified with silanes by both methods, the γ_s^{P} values are lower compared to the unmodified fabrics. The values of the polar components are higher than the dispersive component values only for the fabrics modified with polydopamine. Unmodified fabrics have a smooth surface with numerous longitudinal cracks and fibrils. The surface is chemically inert, and there are no active functional groups. After application of polydopamine, an increase in γ_s^{P} in relation to γ_s^{d} is caused by an increase in the polarity of the polydopamine surface due to the presence of active functional groups on the fiber surface: –OH, –NH$_2$, or –NH (Figure 4).

3.4. The Durability before and after UV Irradiation

3.4.1. Specific Strength

The specific strength of the *m*Ar fabric (8.6 N/tex) (Figure 10a) is 4.5 times lower than that of the *p*Ar fabric, which does not change for *m*Ar/RF/PD/5Ag and *m*Ar/RF/PD/5Ag/S. For *m*Ar/RF/PD/5Ag+S, the specific strength is higher than for the *m*Ar fabric by about 12%. In the case of the modified *p*Ar fabrics, a decrease by about 19%, 14%, and 17% is observed for *p*Ar/RF/PD/5Ag, *p*Ar/RF/PD/5Ag+S, and *p*Ar/RF/PD/5Ag/S, respectively, compared to the unmodified fabric. The higher specific strength values after the application of the Ag+S mixture compared with the AgNWs modified fabrics are probably due to the stronger cross-linking of AgNWs with silanes, which occurred both on the surface of the fabric fibers and in the spaces between the fibers. For the layer-by-layer method, the cross-linking between the AgNWs and the silanes layer takes place only on the surface. The layer-by-layer modification does not change the specific strength of the fabrics in relation to the fabrics with AgNWs.

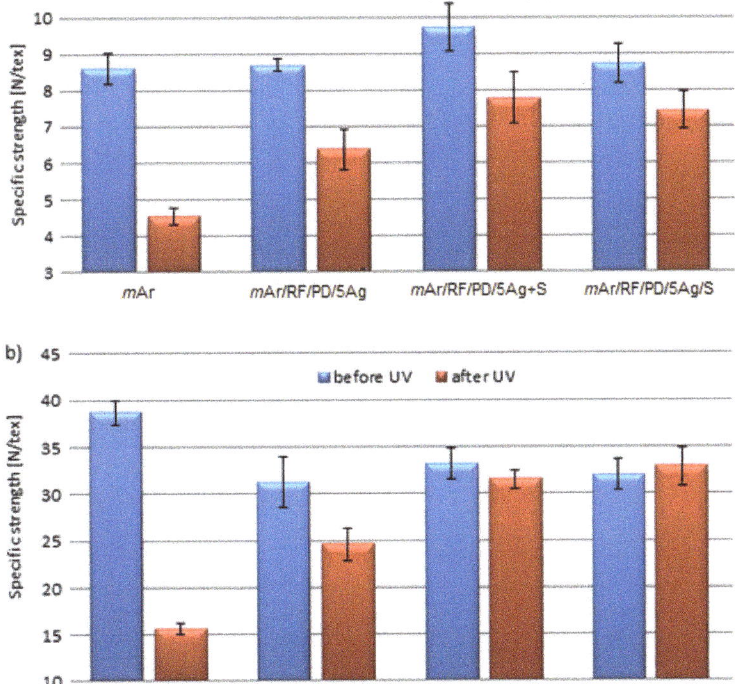

Figure 10. Specific strength of unmodified and modified *m*Ar (**a**) and *p*Ar (**b**) fabrics before and after UV irradiation (365 nm, 96 h).

After 96 h of UV irradiation, an almost 50% decrease in the specific strength is observed for the *m*Ar fabric (Figure 10a). The specific strength is higher by 40%, 71%, and 63% for *m*Ar/RF/PD/5Ag, *m*Ar/RF/PD/5Ag+S, and *m*Ar/RF/PD/5Ag/S, respectively, compared with the *m*Ar fabric after UV irradiation. For the unmodified *p*Ar fabric, there is a 60% decrease in the specific strength. For *p*Ar/RF/PD/5Ag, *p*Ar/RF/PD/5Ag+S, and *p*Ar/RF/PD/5Ag/S (Figure 10b), their specific strength values are 58%, 102%, and 111% higher than for the UV irradiated *p*Ar fabric, respectively. The significantly higher specific strength values after the AgNWs and silanes mixture and layer-by-layer modification is related to the reflection of UV radiation by silanes, which protects the AgNWs and aramid from UV degradation. Shi et al. [28] also observed the effect of the silane layer on the reflectance of UV radiation for a cotton fabric modified with graphene oxide and a KH570 silane layer. Tragoonwichian et al. [29] modified a cotton fabric using vinyletriethoxysilane, which significantly reduced the transmittance of UV radiation through the fabric.

3.4.2. Abrasion Resistance

The *m*Ar fabric is abrasion resistant. After 100,000 cycles (Figure 11a), no breakage of the threads is observed (Figure S1a, Supplementary Materials). For the *p*Ar fabric, thread abrasion (Figure S1b, Supplementary Materials) occurs after 5000 cycles (Figure 11b), probably due to the less ordered structure and lower degree of crystallinity of *m*Ar than of *p*Ar. For the *m*Ar/RF/PD/5Ag and *m*Ar/RF/PD/5Ag+S fabric, a decrease in the number of abrasion cycles by 40,000 compared to the *m*Ar fabric is observed, which proves that AgNWs reduce the abrasion resistance of *meta*-aramid. In the case of the *para*-aramid fabrics, the application of AgNWs and a mixture of AgNWs and silanes improves the abrasion resistance and increases the number of abrasion cycles (Figure 11b) by 7000 and 75,000, respectively. For the layer-by-layer method, no thread damage is observed for both fabrics after 100,000 abrasion cycles.

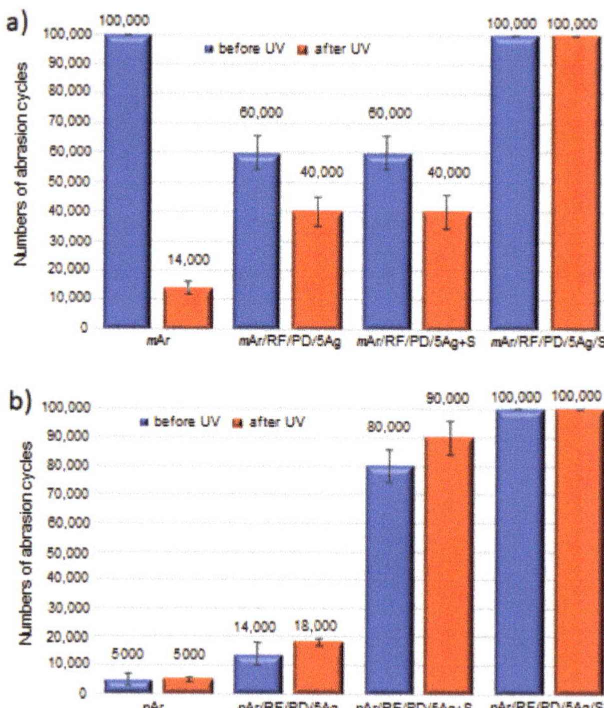

Figure 11. Results of the Martindale's abrasion test for unmodified and modified *m*Ar (**a**) and *p*Ar (**b**) fabrics before and after UV irradiation (365 nm, 96 h).

After UV exposure, the abrasion resistance of the mAr fabric decreased, the breakage of the threads is observed (Figure S1a, Supplementary Materials), and the number of abrasion cycles decreases by a factor of seven. For the *m*Ar/RF/PD/5Ag and *m*Ar/RF/PD/5Ag+S fabric, the number of abrasion cycles is three times higher than that of the *m*Ar fabric after UV irradiation. For the *p*Ar fabric, the number of abrasion cycles does not change after UV exposure. For the modified fabrics, it is almost four times and 18 times higher for *p*Ar/RF/PD/5Ag and *p*Ar/RF/PD/5Ag+S, respectively, compared to the UV irradiated *p*Ar fabrics. For the layer-by-layer method, after UV irradiation, no breakage of the fibers is observed after 100,000 cycles for both types of aramids, as before UV exposure (Figure S1b, Supplementary Materials).

3.5. Conductive Properties

The unmodified *m*Ar and *p*Ar fabrics are non-conductive, and their electrical surface resistance (R_s) is 1.30×10^{12} Ω and 1.26×10^{12} Ω (Figure 12a), respectively. For both aramid fabrics after the 5-fold AgNWs application, a significant decrease in R_s values by 10 orders of magnitude is observed. For both aramid fabrics modified with silanes by the mixture and the layer-by-layer method, a decrease in the R_s value by eight and nine orders of magnitude, respectively, is observed compared with the reference fabrics.

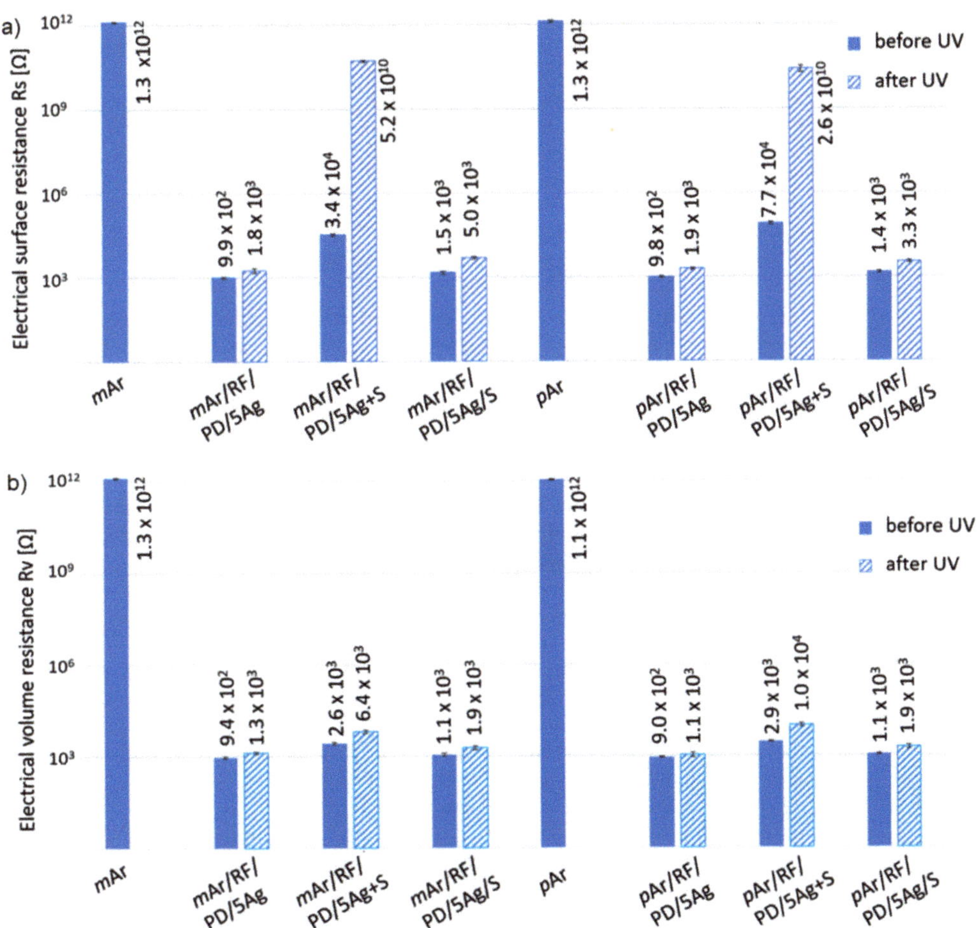

Figure 12. Electrical surface resistance (R_s) (**a**) and electrical volume resistance (R_v) (**b**) of unmodified and modified *m*Ar and *p*Ar fabrics before and after UV irradiation (365 nm, 96 h).

The electrical volume resistance (R_v) value is 1.30×10^{12} Ω and 1.06×10^{12} Ω (Figure 12b) for the *m*Ar and *p*Ar fabrics, respectively. After the 5-fold AgNW application, the R_v value decreases by 10 orders of magnitude for both fabrics. After the AgNWs and silanes mixture and layer-by-layer modification, the R_v values are lower than those of the unmodified fabrics by nine orders of magnitude.

The slightly higher values of R_s and R_v for the fabrics modified with silanes by the mixture and layer-by-layer method compared to the fabrics with only AgNWs are due to the presence of silanes, which, on aramid fabrics, have higher R_s and R_v values of 5.98×10^7 Ω and 3.77×10^7 Ω, respectively, for the *m*Ar fabric and 3.30×10^8 Ω and 6.91×10^7 Ω, respectively, for the *p*Ar fabric. Silanes can be a barrier to the flow of electrons through AgNWs. Moreover, higher values of both resistances are found for the mixture method than for the layer-by-layer method, which may be due to the AgNWs being covered by silanes, which limits their direct connection with each other (which can be seen in the SEM images, Figures 7c and 8c). On the other hand, for the layer-by-layer method, fibers that protrude above the silanes layer are densely covered with AgNWs, which create conductive paths. Under the silanes layer, both on the surface and between the fibers, there are AgNWs that connect with each other, which ensures a lower R_v value compared to the fabrics modified with the mixture method.

Moreover, the lower values of R_v in relation to R_s for all modified fabrics may be due to the AgNWs filling the spaces between the fibers.

After UV irradiation, the R_s values increase by one order of magnitude for both the AgNWs modified fabrics (Figure 12a). This increase may be caused by the degradation of AgNWs, which is visible in the SEM images as silver precipitates on the surface of the nanowires (Figures 7b and 8b). For the mixture method, the R_s values increase by six orders of magnitude and the fabric becomes non-conductive. For the layer-by-layer method, the R_s value has the same order of magnitude. These differences are related to the different thicknesses of the silanes layer. For the layer-by-layer method, the layer is thicker and protects the AgNWs from the UV radiation. In the case of the mixture method, the silanes coating that covers the AgNWs is thin and does not reflect the UV radiation as efficiently. For fabrics modified with the AgNWs and silanes mixture, the coating is discontinuous and damaged with numerous cavities and cracks (Figures 7c and 8c). This causes interruption percolating paths formed by the AgNWs, which prevent the flow of electrons.

The UV radiation causes an increase in the R_v values for the AgNWs modified fabrics by one order of magnitude (Figure 12b). For the layer-by-layer method, the value only slightly increases. The R_v values of *m*Ar/RF/PD/5Ag+S increase but have the same order of magnitude. For *p*Ar/RF/PD/5Ag+S, an increase by one order of magnitude is observed. The lower increase in R_v compared to R_s after irradiation is due to the fact that AgNWs located between the fibers inside the fabric are less exposed to UV radiation than the AgNWs on its surface.

4. Conclusions

Multifunctional *meta-* and *para*-aramid fabrics modified with AgNWs and a mixture of silanes, APTES, and DEDMS, were obtained. The aim of the studies was to protect against UV radiation and improve the mechanical, electrical, and hydrophobic properties of modified aramid fabrics. New functionalization methods, a one-step method (mixture) with AgNWs dispersed in the silanes mixture, and a two-step method (layer-by-layer) in which the silanes mixture was applied on the previously deposited AgNWs layer, were developed.

Fabrics were pre-treated in a low-pressure air RF plasma, and a subsequent poly-dopamine coating was applied.

The modified fabrics acquired hydrophobic properties. The water contact angle for the *meta-* and *para*-aramid fabric was 125° and 120°, respectively, for the mixture and 112° and 114°, respectively, for the layer-by-layer method. The surface free energy for the AgNWs and silanes modified fabrics is lower by about 30% for the *meta*-aramid fabric and 50% for the *para*-aramid fabric compared with the reference fabrics.

Better resistance to UV radiation was achieved for the layer-by-layer method. The UV radiation caused no changes in the layer-by-layer coating, while for the mixture coating, there was damage, and some coating fragments were detached. The UV radiation caused a significant decrease in the specific strength of the reference fabrics by about 50% and 60% for the *meta-* and *para*-aramid fabric, respectively, while for the mixture and layer-by-layer method, the specific strength values were 71% and 63% higher for the *meta*-aramid fabric, respectively, and 102% and 110% higher for the *para*-aramid fabric, respectively. The layer-by-layer modified fabrics were the most abrasion resistant; before and after UV radiation, no thread damage was observed after 100,000 abrasion cycles. The electrical surface resistance for the layer-by-layer method was nine orders of magnitude lower than the reference fabrics, and for the mixture method, it was eight orders of magnitude lower. After UV irradiation, the R_s did not change for the layer-by-layer method and increased by six orders of magnitude for the mixture modified fabrics, which became non-conductive. We selected the layer-by-layer method as the most effective for aramid fabric modification; it results in great surface, electrical, and mechanical properties and better resistance to UV radiation. It protects aramid fabrics from degradation and the deterioration of their functional properties.

Supplementary Materials: The following supporting information can be downloaded at: https://www.mdpi.com/article/10.3390/molecules27061952/s1, Figure S1: Images of (a) *m*Ar and (b) *p*Ar fabrics surfaces, before and after UV radiation (365 nm, 96 h), before and after abrasion cycles.

Author Contributions: Conceptualization, A.N., M.C. and G.C.; formal analysis, A.N., M.C. and G.C.; visualization, A.N.; funding acquisition, M.C. and G.C.; investigation, A.N. and M.C.; methodology, A.N., A.B.-K., K.R.-S., I.J. and G.C.; project administration, M.C. and G.C.; writing—original draft, A.N.; writing—review and editing, A.N., M.C. and G.C. All authors have read and agreed to the published version of the manuscript.

Funding: The research was funded by the National Science Center, project grant: 2018/29/B/ST8/02016 and the internal project BZT 0166/2020-2021. The research was carried out within the National Science Center, on the apparatus purchased in the Key Project—POIG.01.03.01-00-004/08 Functional nano and micro textile materials—NANOMITEX and WND-RPLD. 03.01.00-001/09.

Institutional Review Board Statement: Not applicable.

Informed Consent Statement: Not applicable.

Data Availability Statement: Not applicable.

Conflicts of Interest: The authors declare no conflict of interest.

Sample Availability: Samples of the compounds are not available from the authors.

References

1. Zhang, H.; Liang, G.; Gu, A.; Yuan, L. Facile preparation of hyperbranched polysiloxane-grafted aramid fibers with simulta-neously improved UV resistance, surface activity, and thermal and mechanical properties. *Ind. Eng. Chem. Res.* **2014**, *53*, 2684–2696. [CrossRef]
2. Wang, H.; Xie, H.; Hu, Z.; Wu, D.; Chen, P. The influence of UV radiation and moisture on the mechanical properties and micro-structure of single Kevlar fibre using optical methods. *Polym. Degrad. Stab.* **2012**, *97*, 1755–1761. [CrossRef]
3. Zhu, J.; Yuan, L.; Guan, Q.; Liang, G.; Gu, A. A novel strategy of fabricating high performance UV-resistant aramid fibers with simultaneously improved surface activity, thermal and mechanical properties through building polydopamine and graphene oxide bi-layer coatings. *Chem. Eng. J.* **2017**, *310*, 134–147. [CrossRef]
4. Foksowicz-Flaczyk, J.; Walentowska, J.; Przybylak, M.; Maciejewski, H. Multifunctional durable properties of textile materials modified by biocidal agents in the sol-gel process. *Surf. Coat. Technol.* **2016**, *304*, 160–166. [CrossRef]
5. Stoppa, M.; Chiolerio, A. Wearable Electronics and Smart Textiles: A Critical Review. *Sensors* **2014**, *14*, 11957–11992. [CrossRef] [PubMed]
6. Gowri, S.; Almeida, L.; Amorim, T.; Carneiro, N.; Souto, A.P.; Esteves, M.F. Polymer nanocomposites for multifunctional fin-ishing of textiles—A review. *Text. Res. J.* **2014**, *80*, 1290–1306. [CrossRef]
7. Nejman, A.; Kamińska, I.; Jasińska, I.; Celichowski, G.; Cieślak, M. Influence of Low-Pressure RF Plasma Treatment on Aramid Yarns Properties. *Molecules* **2020**, *25*, 3476. [CrossRef]
8. Inagaki, N.; Tasaka, S.; Kawai, H. Surface modification of Kevlar® fiber by a combination of plasma treatment and coupling agent treatment for silicone rubber composite. *J. Adhes. Sci. Technol.* **1992**, *6*, 279–291. [CrossRef]
9. Sun, Z.; Zhou, Y.; Li, W.; Chen, S.; You, S.; Ma, J. Preparation of Silver-Plated Para-Aramid Fiber by Employing Low-Temperature Oxygen Plasma Treatment and Dopamine Functionalization. *Coatings* **2019**, *9*, 599. [CrossRef]
10. Sa, R.; Yan, Y.; Wei, Z.; Zhang, L.; Wang, W.; Tian, M. Surface Modification of Aramid Fibers by Bio-Inspired Poly(dopamine) and Epoxy Functionalized Silane Grafting. *ACS Appl. Mater. Interfaces* **2014**, *6*, 21730–21738. [CrossRef]
11. Chai, D.; Xie, Z.; Wang, Y.; Liu, L.; Yum, Y.-J. Molecular Dynamics Investigation of the Adhesion Mechanism Acting between Dopamine and the Surface of Dopamine-Processed Aramid Fibers. *ACS Appl. Mater. Interfaces* **2014**, *6*, 17974–17984. [CrossRef] [PubMed]

12. Zeng, L.; Liu, X.; Chen, X.; Soutis, C. Surface Modification of Aramid Fibres with Graphene Oxide for Interface Improvement in Composites. *Appl. Compos. Mater.* **2018**, *25*, 843–852. [CrossRef]
13. Lee, H.; Dellatore, S.M.; Miller, W.M.; Messersmith, P.B. Mussel-Inspired Surface Chemistry for Multifunctional Coatings. *Science* **2007**, *318*, 426–430. [CrossRef]
14. Wei, Q.; Zhang, F.; Li, J.; Li, B.; Zhao, C. Oxidant-induced dopamine polymerization for multifunctional coatings. *Polym. Chem.* **2010**, *1*, 1430–1433. [CrossRef]
15. Lee, H.; Scherer, N.F.; Messersmith, P.B. Single-molecule mechanics of mussel adhesion. *Proc. Natl. Acad. Sci. USA* **2006**, *103*, 12999–13003. [CrossRef] [PubMed]
16. Burzio, L.A.; Waite, J.H. Cross-Linking in Adhesive Quinoproteins: Studies with Model Decapeptides. *Biochemistry* **2000**, *39*, 11147–11153. [CrossRef] [PubMed]
17. Sever, M.J.; Weisser, J.T.; Monahan, J.; Srinivasan, S.; Wilker, J.J. Metal-mediated cross-linking in the generation of a ma-rine-mussel adhesive. *Angew. Chem. Int. Ed.* **2004**, *43*, 448–450. [CrossRef]
18. Yu, M.; Hwang, J.; Deming, T.J. Role of l-3,4-Dihydroxyphenylalanine in Mussel Adhesive Proteins. *J. Am. Chem. Soc.* **1999**, *121*, 5825–5826. [CrossRef]
19. Gu, T.; Zhu, D.; Lu, S. Surface Functionalization of Silver-Coated Aramid Fiber. *Polym. Sci. Ser. A* **2020**, *62*, 196–204. [CrossRef]
20. Wang, W.; Li, R.; Tian, M.; Liu, L.; Zou, H.; Zhao, X.; Zhang, L. Surface Silverized Meta-Aramid Fibers Prepared by Bio-inspired Poly(dopamine) Functionalization. *ACS Appl. Mater. Interfaces* **2013**, *5*, 2062–2069. [CrossRef]
21. Liu, H.; Zhu, L.; Xue, J.; Hao, L.; Li, J.; He, Y.; Cheng, B. A Novel Two-Step Method for Fabricating Silver Plating Cotton Fabrics. *J. Nanomater.* **2016**, *2016*, 2375836. [CrossRef]
22. Sun, Y.; Gates, B.; Mayers, B.; Xia, Y. Crystalline Silver Nanowires by Soft Solution Processing. *Nano. Lett.* **2002**, *2*, 165–168. [CrossRef]
23. Wang, Z.; Liu, J.; Chen, X.; Wan, J.; Qian, Y. A Simple Hydrothermal Route to Large-Scale Synthesis of Uniform Silver Nanowires. *Chem.—Eur. J.* **2004**, *11*, 160–163. [CrossRef] [PubMed]
24. James, S.; Robinson, A.; Arnold, J.; Worsley, D. The effects of humidity on photodegradation of poly(vinyl chloride) and polyethylene as measured by the CO_2 evolution rate. *Polym. Degrad. Stab.* **2013**, *98*, 508–513. [CrossRef]
25. Mather, R.R.; Wardman, R.H. *The Chemistry of Textile Fibers*; RSC Publishing: Cambridge, UK, 2011.
26. Lin, M.C.; Chu, C.J.; Tsai, L.C.; Lin, H.Y.; Wu, C.S.; Wu, Y.P.; Shieh, D.B.; Su, Y.W.; Chen, C.D. Control and Detection of Organosilane Polarization on Nanowire Field-Effect Transistors. *Nano Lett.* **2007**, *7*, 3656–3661. [CrossRef]
27. Chiavari, C.; Balbo, A.; Zanotto, F.; Vassura, I.; Bignozzi, M.C.; Monticelli, C. Organosilane coatings applied on bronze: In-fluence of UV radiation and thermal cycles on the protectiveness. *Progr. Org. Coat.* **2015**, *82*, 91–100. [CrossRef]
28. Shi, F.; Xu, J.; Zhang, Z. Study on UV-protection and hydrophobic properties of cotton fabric functionalized by graphene oxide and silane coupling agent. *Pigment Resin Technol.* **2019**, *48*, 237–242. [CrossRef]
29. Tragoonwichian, S.; O'Rear, E.A.; Yanumet, N. Grafting polymerization of 2-[3-(2h-benzotriazol-2-yl)-4-hydroxyphenyl]ethyl methacrylate on vinyltriethoxysilane-treated cotton for the preparation of ultraviolet-protective fabrics. *J. Appl. Polym. Sci.* **2009**, *114*, 62–69. [CrossRef]
30. Roe, B.; Zhang, X. Durable Hydrophobic Textile Fabric Finishing Using Silica Nanoparticles and Mixed Silanes. *Text. Res. J.* **2009**, *79*, 1115–1122. [CrossRef]
31. Zeng, X.; Xu, G.; Gao, Y.; An, Y. Surface Wettability of (3-Aminopropyl)triethoxysilane Self-Assembled Monolayers. *J. Phys. Chem. B* **2010**, *115*, 450–454. [CrossRef]
32. Zhang, Y.; Shen, Q.; Li, X.; Xie, H.; Nie, C. Facile synthesis of ternary flexible silica aerogels with coarsened skeleton for oil–water separation. *RSC Adv.* **2020**, *10*, 42297–42304. [CrossRef]
33. Souza, K.G.D.S.; Cotting, F.; Aoki, I.V.; Amado, F.D.R.; Capelossi, V.R. Study of the wettability and the corrosion protection of the hybrid silane (3-aminopropyl) triethoxysilane (APTES) and (3-glycidyloxypropyl) trimethoxysilane (GPTMS) film on galvannealed steel. *Matéria* **2020**, *25*. [CrossRef]
34. Hasanzadeh, M.; Far, H.S.; Haji, A.; Rosace, G. Facile fabrication of breathable and superhydrophobic fabric based on silica na-noparticles and amino-modified polydimethylsiloxane. *Preprints* **2020**, *257*, 5491–5498.
35. Liu, H.; Lee, Y.-Y.; Norsten, T.B.; Chong, K. In situ formation of anti-bacterial silver nanoparticles on cotton textiles. *J. Ind. Text.* **2013**, *44*, 198–210. [CrossRef]
36. Giesz, P.; Mackiewicz, E.; Nejman, A.; Celichowski, G.; Cieślak, M. Investigation on functionalization of cotton and viscose fabrics with AgNWs. *Cellulose* **2016**, *24*, 409–422. [CrossRef]
37. Schmidt, H.; Witkowska, B.; Kamińska, I.; Twarowska-Schmidt, K.; Wierus, K.; Puchowicz, D. Comparison of the rates of polypropylene fibre degradation caused by artificial light and sunlight. *Fib. Tex. East. Eur.* **2011**, *4*, 53–58.
38. Owens, D.K.; Wendt, R.C. Estimation of the surface free energy of polymers. *J. Appl. Polym. Sci.* **1969**, *13*, 1741–1747. [CrossRef]
39. Cieslak, M.; Puchowicz, D.; Schmidt, H. Evaluation of the possibility of using surface free energy study to design protective fabrics. *Text. Res. J.* **2011**, *82*, 1177–1189. [CrossRef]
40. Giesz, P.; Celichowski, G.; Puchowicz, D.; Kamińska, I.; Grobelny, J.; Batory, D.; Cieślak, M. Microwave-assisted TiO_2: Anatase formation on cotton and viscose fabric surfaces. *Cellulose* **2016**, *23*, 2143–2159. [CrossRef]
41. Schmidt, H.; Cieślak, M. Concrete with carpet recyclates: Suitability assessment by surface energy evaluation. *Waste Manag.* **2008**, *28*, 1182–1187. [CrossRef]

42. Sabdin, S.; Azmi, M.A.M.; Badruzaman, N.A.; Makmon, F.Z.; Aziz, A.A.; Said, N.A.M. Effect of APTES Percentage towards Reduced Graphene Oxide Screen Printed Electrode Surface for Biosensor Application. *Mater. Today Proc.* **2019**, *19*, 1183–1188. [CrossRef]
43. Jiang, J.; Zhu, L.; Zhu, L.; Zhang, H.; Zhu, B.; Xu, Y. Antifouling and Antimicrobial Polymer Membranes Based on Bioinspired Polydopamine and Strong Hydrogen-Bonded Poly(N-vinyl pyrrolidone). *ACS Appl. Mater. Interfaces* **2013**, *5*, 12895–12904. [CrossRef] [PubMed]
44. Chen, S.; Zhang, L.; Sun, M.; Zhang, X.; Chen, W. Surface modification of polypropylene nonwoven fabrics by grafting of polydopamine. *Adv. Polym. Technol.* **2018**, *37*, 3519–3528. [CrossRef]
45. Zhang, Y.; Bai, S.; Chen, T.; Yang, H.; Guo, X. Facile preparation of flexible and highly stable graphene oxide-silver nanowire hybrid transparent conductive electrode. *Mater. Res. Express* **2019**, *7*, 016413. [CrossRef]

Article

Antibacterial Properties of Non-Modified Wool, Determined and Discussed in Relation to ISO 20645:2004 Standard

Tomislav Ivankovic [1,*], Antonija Rajic [1], Sanja Ercegovic Razic [2], Sabine Rolland du Roscoat [3] and Zenun Skenderi [2]

1 Department of Biology, Faculty of Science, University of Zagreb, 10000 Zagreb, Croatia; tona31@gmail.com
2 Department of Materials, Fibres and Textile Testing, Faculty of Textile Technology, University of Zagreb, 10000 Zagreb, Croatia; sanja.ercegovic@ttf.unizg.hr (S.E.R.); zenun.skenderi@ttf.unizg.hr (Z.S.)
3 Laboratoire Sols, Solides, Structures et Risques (3SR), UMR 5521, Université Grenoble Alpes, CNRS G-INP, 38000 Grenoble, France; sabine.rolland-du-roscoat@univ-grenoble-alpes.fr
* Correspondence: tomislav.ivankovic@biol.pmf.hr

Abstract: Wool is considered to possibly exhibit antibacterial properties due to the ability of wool clothing to reduce the build-up of odor, which arises from the microbial activity of skin microbiota. Indeed, when tested with a widely used agar diffusion plate test method, even wool or other textiles not treated with any antimicrobial agent can be interpreted to show certain antibacterial effects due to the lack of growth under the specimen, as instructed in ISO 20645:2004 standard. Therefore, we analyzed in detail what happens to bacterial cells in contact with untreated wool and cotton fabric placed on inoculated agar plates by counting viable cells attached to the specimens after 1 and 24 h of contact. All wool and several cotton samples showed no growth under the specimen. Nevertheless, it was shown without a doubt that neither textile material kills bacteria or inhibits cell multiplication. A reasonable explanation is that bacterial cells firmly attach to wool fibers forming a biofilm during multiplication. When the specimen was lifted off the nutrient agar surface, the cells in the form of biofilm remained attached to the wool fibers, removing the biomass and resulting in a clear, no growth zone underneath it. By imaging the textile specimens with X-ray microtomography, we concluded that the degree of attachment could be dependent on surface topography. The results indicate that certain textiles, in this case, wool, could exhibit antibacterial properties by removing excess bacteria that grow on the textile/skin interface when taken off the body.

Keywords: textile; cotton; ISO standards; antimicrobial; agar diffusion

check for **updates**

Citation: Ivankovic, T.; Rajic, A.; Ercegovic Razic, S.; Rolland du Roscoat, S.; Skenderi, Z. Antibacterial Properties of Non-Modified Wool, Determined and Discussed in Relation to ISO 20645:2004 Standard. *Molecules* **2022**, 27, 1876. https://doi.org/10.3390/molecules27061876

Academic Editor: Baljinder Kandola

Received: 1 February 2022
Accepted: 9 March 2022
Published: 14 March 2022

Publisher's Note: MDPI stays neutral with regard to jurisdictional claims in published maps and institutional affiliations.

1. Introduction

The possibility that wool has antibacterial properties comes from the ability of wool clothing to reduce/resist the onset of odor build-up, and odor is primarily considered to originate from microbial activity [1–3]. Caven et al. [1] suggested three possible explanations for the supposed antibacterial properties of wool. First, the complex wool fiber composed of epicuticle, lipid monolayer, and the cortex has an antibacterial effect, as suggested by Johnson et al. [3]. Nowadays, this statement does not seem to be correct, as several studies have undoubtedly shown that untreated wool itself does not exhibit bactericidal or bacteriostatic properties [1,4,5]. These studies used the absorption method (i.e., ISO 20743:2013) and compared the number of bacteria inoculated with liquid nutrient media onto the fiber specimen and the number of bacteria on the specimen after a certain period of time, showing that bacteria either remained viable or multiplied on the wool fibers. In one of our previous studies, the number of bacteria inoculated, according to the absorption method, on untreated wool fabric increased by four log values after 24 h of contact [6].

The second explanation would be that wool bonds or adsorbs odorous fumes without actually inhibiting bacterial growth, for which there is strong evidence [7–9].

The third possibility is that wool fibers' hydrophobic surface and specific microstructure create a microclimate unfavorable for bacterial growth [1]. This third possibility was the object of our investigation. To simulate the real-life conditions where wool is in direct contact with skin covered with normal microbiota, we tested scoured plain weave woolen fabric by using the standard method ISO 20645:2004 "Textile fabrics—Determination of antibacterial activity—Agar diffusion plate test." In this method, the textile specimen is placed on top of the inoculated agar, and as the antimicrobial agent diffuses into the agar, it either kills or stops bacterial cells from multiplying, giving a clear zone of "no bacterial growth" around the specimen. However, according to the method instructions, even the observation of no growth under the specimen can be classified as a "good effect" (the term "good effect" is a direct quote from the document).

However, there is a methodological issue with the agar diffusion test; the ISO 20645:2004 is suitable only to test the textiles treated with an antimicrobial agent compared with the control, untreated specimens [10,11]. However, earlier, we observed that untreated wool and some other untreated textiles also show "no growth" under the specimen when tested by the agar diffusion test [6]. Thus, can this be classified as an antibacterial effect?

To try and provide an answer, we further investigated the results of agar diffusion testing by determining what is happening with the bacteria in contact with textile samples. This was undertaken by counting the viable cells on the specimens after 1 h and 24 h of contact of the textile specimen with agar inoculated with bacteria (Figure 1).

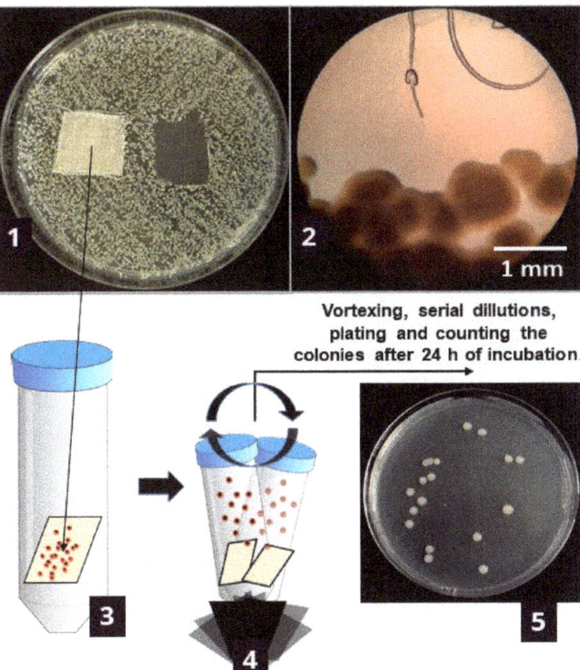

Figure 1. Experimental setup for determining the number of viable bacteria attached to the samples during the contact with inoculated agar. Initially, the bacteria are inoculated across the nutrient agar plate, and specimens are placed onto the plate. After the incubation (37 °C/24 h), the growth under the specimen was determined visually (**1**) and under the microscope (**2**). The textile specimen was transferred to 20 mL of sterile saline solution and vortexed five times for 5-sec bursts (**3**) to detach the bacterial cells from the specimen. The supernatant was serially diluted up to 10^{-7}; plated and grown colonies were counted after the incubation at 37 °C/24 h (**4**). The number of bacteria was reported as log CFU·cm^{-2} of the specimen (**5**).

By doing this, we wanted to unravel the methodological issue of interpreting the "no growth" under the sample (instructed by ISO 20645:2004 method) and to define whether non-modified wool and other textiles adsorb the bacteria, kill the bacteria, or stop their growth, or do not have any effect on bacterial cells at all.

2. Results and Discussion

2.1. Agar Diffusion Test

We assessed the antibacterial efficacy of three types of tested textiles, namely wool fabric, the cotton of a standard laboratory coat, and cotton of standard sterile compressed gauze, toward several bacterial species (Table 1). Bacterial species that were used in the present research were chosen to represent diversity regarding their metabolic and morphologic characteristics. The *Staphylococcus aureus* and *Klebsiella pneumoniae* are respectively Gram-positive and Gram-negative bacteria are listed as test organisms in ISO 20645:2004 standard. In our test battery, we added another Gram-positive bacterium, *Bacillus cereus*, which produces endospores. Finally, three strains of Gram-negative opportunistic pathogenic bacteria *Acinetobacter baumannii* were tested as well, an ATCC type strain and two hospital isolates resistant to multiple antibiotics.

Table 1. Assessment of antibacterial efficacy of textile materials determined with the ISO 20645:2004 "Agar diffusion plate test". The material/bacteria combination that yielded "good effect" is shaded.

Bacteria	Sample	Growth Under the Specimen	Assessment of Antibacterial Efficacy Adopted from ISO 20645:2004
S. aureus ATCC 25923	Wool	none/slight	good effect/limit of efficacy
	Cotton	heavy	insufficient effect
	Gauze	slight/moderate	limit of efficacy/insufficient effect
K. pneumoniae ATCC 11296	Wool	none	good effect
	Cotton	none/slight	good effect/limit of efficacy
	Gauze	slight	limit of efficacy
B. cereus LBK 4080	Wool	none	good effect
	Cotton	none/slight	good effect/limit of efficacy
	Gauze	slight/moderate	limit of efficacy/insufficient effect
A. baumannii ATCC 11296	Wool	none	good effect
	Cotton	none	good effect
	Gauze	slight	limit of efficacy
A. baumannii HI1	Wool	none	good effect
	Cotton	none/slight	good effect/limit of efficacy
	Gauze	slight	limit of efficacy
A. baumannii HI2	Wool	none	good effect
	Cotton	none	good effect
	Gauze	slight	limit of efficacy

The results were interpreted by adopting the ISO 20645:2004 assessment and observing the growth under the textile specimen (Table 1). "No growth" under the specimen (presented in Figure 2b) was assessed as good antibacterial effect, "slight growth" was assessed as a limit of efficacy whilst "moderate" (Figure 2c) or "heavy growth" (Figure 2d) was assessed as no effect. Aquacell®, a wound dressing doped with silver, was used as the positive control, showing no growth under the specimen along with an inhibition zone around the sample (Figure 2a), demonstrating antibacterial activity of ionic silver.

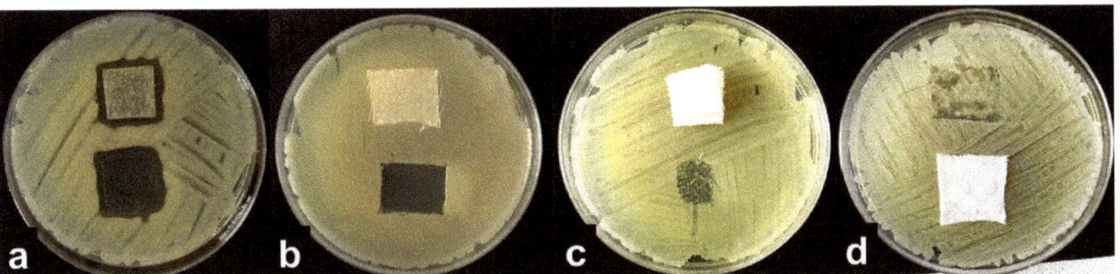

Figure 2. Interpretation of the agar diffusion test results based on ISO 20645:2004 standard: inhibition zone and no growth under the Aquacell® specimen (**a**) in the experiment with *A. baumannii*. No inhibition zone and no growth under the wool fabric specimen (**b**) in the experiment with *B. cereus*. No inhibition zone and slight/moderate growth under the gauze specimen (**c**) in the experiment with *K. pneumoniae*. No inhibition zone and heavy growth under the cotton specimen (**d**) in the experiment with *S. aureus*.

Comparing the three materials, the best antibacterial effect was exhibited by wool fabric (Table 1). Wool showed good antibacterial effect towards all tested bacteria except *S. aureus*, where it was at the limit of efficacy. Cotton was at the limit of efficacy towards all of the bacteria, again except *S. aureus* where it had insufficient effect, and two strains of *A. baumannii* where it showed good effect. Compressed gauze exhibited either insufficient effect or limit of efficacy. Comparing different bacteria, the results were similar for each material (Table 1), suggesting that the specific textile affects all bacteria in the same manner, again exempting *S. aureus*, which seemed to have the strongest growth under the specimen, regardless of the material. To summarise, out of the six different bacteria we tested, the wool exhibited good antibacterial effect toward five of them, the cotton toward two of them, and compressed gauze towards none (Table 1).

2.2. Comparison to Literature Data

The comparison of our results with literature data was somewhat challenging. A standardized method for antibacterial testing of textile samples resembling ours is perhaps the AATCC 147 Standard [12], where bacteria are inoculated on top of the agar plates as parallel streaks. The textile specimen is placed on the agar surface, and the inhibition zone around the specimen is monitored. Therefore, this method also enables direct contact of bacterial cells with the textile, assessment of growth under the sample, and can be used for textiles without diffusible agents [10]. We were able to find only one study reporting growth under the untreated wool sample; Liu et al. [13] tested capsaicin-coated wool fabric and reported "heavy growth" under control, untreated wool fabric, indicating no antibacterial effect. Another case of a very similar experimental setup was reported in Gomes et al. [14], where cotton modified with chitosan was tested by the JIS L 1902-Halo standard method. This method is very similar to ISO 20645:2004 and is also essentially a pour plate method; the difference is that "ISO" demands two layers of agar, the bottom layer clean and the upper layer inoculated with bacteria, while the "Halo method" demands only one layer of agar inoculated with bacterial culture. However, Gomes et al. [14] inoculated bacterial suspension on top of the agar layer, making it identical to our setup. They report no antibacterial effect on control untreated cotton, but only the "Halo zone", the inhibition zone around the specimen, was monitored. There is no mention of bacterial growth under the specimen, preventing comparison to our results. A modified Kirby—Bauer test, a non-standardized method for antibacterial testing of textiles, is identical to the experimental setup we were using. However, we could not find any mention that growth under any textile, let alone wool, was monitored and reported, only the inhibition zone [15–17].

What is clear from the available literature is that wool itself does not exhibit bactericidal properties. Using the standard pour plate method, Pollini et al. [18] tested wool treated with silver and reported no antibacterial effect on untreated control samples. The same was reported in a couple of other studies [4,19]. From the aforementioned results, it is clear that the wool itself does not release any antibacterial agent whatsoever. However, as already mentioned, it also does not exhibit antibacterial activity when in contact with bacterial cells, as shown using the standard absorption method [1,4–6].

2.3. Viable Cell Count

As we established that there was no bacterial growth under the majority of the wool specimens and most of the cotton specimens, and wool, according to literature, is not bactericidal itself. The question was what happened to the bacterial cells; were they inhibited to grow, attached to the textile and stopped multiplying or were they attached to the textile and multiplying? To answer this question, we determined the number of bacterial cells on the specimens after 1 h and 24 h of contact (specimens being placed onto an inoculated nutrient agar surface).

All the bacterial species and all the textiles showed the same trend, meaning the bacteria attached to the materials during 1 h of contact and continued to multiply, as indicated by a significant ($p < 0.05$) increase in the bacterial numbers after 24 h of contact (Figures 3–5). The same trend was in both experimental setups when the starting concentration of bacteria was lower (log 5 CFU·mL^{-1}) or higher (log 8 CFU·mL^{-1}). As expected, when initially fewer bacteria were inoculated on agar plates, the lower the number of bacteria attached after 1 h (Figures 3–5). After 24 h the number of attached bacteria was the same, regardless of whether initially we inoculated a 1000-fold lower or higher concentration of bacteria (Figures 3–5). Such results suggest that bacteria have initially attached and then multiplied until reaching some maximum amount, determined not by the number of bacteria initially attached to the textile but by the experimental setup (type of nutrient media, temperature, and time of incubation).

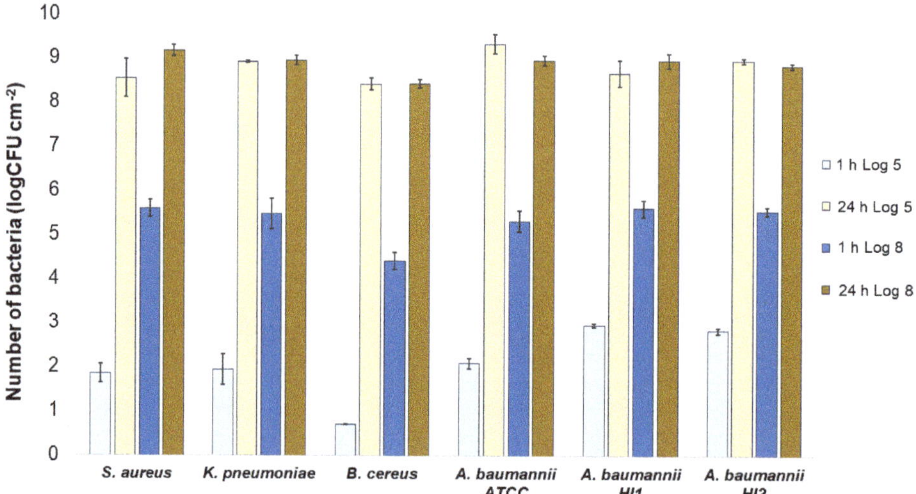

Figure 3. Numbers of viable bacteria that were adsorbed onto wool specimens after 1 h and 24 h of contact in the agar diffusion test. Two experimental setups, one in which the inoculum concentration was low (log 5 = the concentration was ~10^5 CFU·mL^{-1}) and the other in which the inoculum concentration was high (log 8 = the concentration was ~10^8 CFU·mL^{-1}). The inoculum was spread across the surface of the nutrient agar plates before the placement of specimens and incubation.

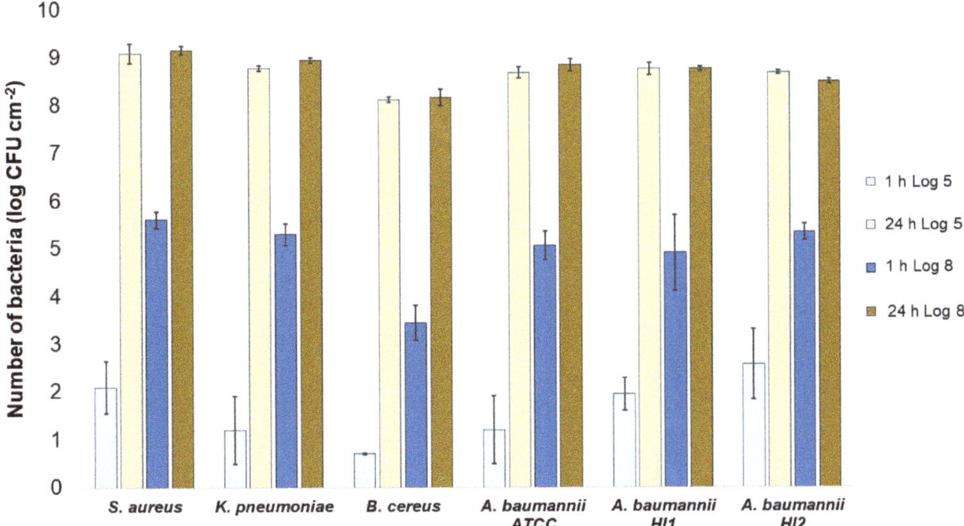

Figure 4. Numbers of viable bacteria that were adsorbed onto cotton specimens after 1 h and 24 h of contact in the agar diffusion test. Two experimental setups, one in which the inoculum concentration was low (log 5 = the concentration was ~10^5 CFU·mL^{-1}) and the other in which the inoculum concentration was high (log 8 = the concentration was ~10^8 CFU·mL^{-1}).

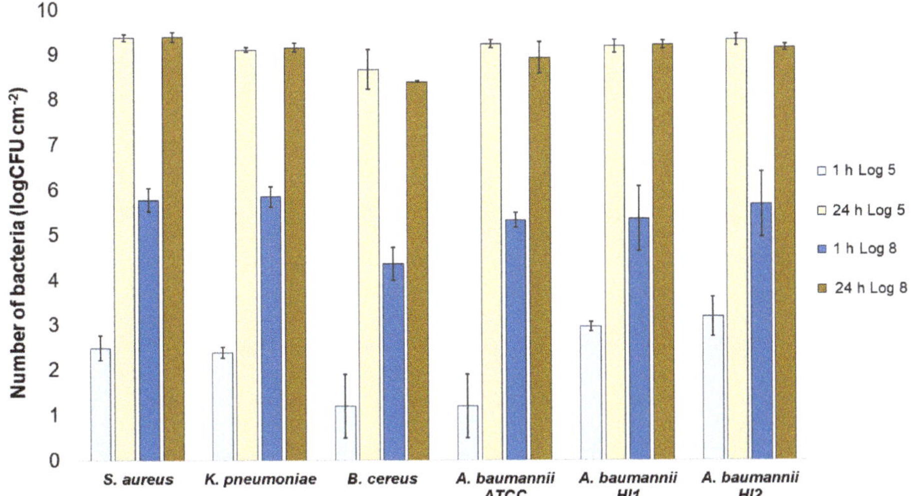

Figure 5. Numbers of viable bacteria that were adsorbed onto compressed gauze specimens after 1 h and 24 h of contact in the agar diffusion test. Two experimental setups, one in which the inoculum concentration was low (log 5 = the concentration was ~10^5 CFU·mL^{-1}) and the other in which the inoculum concentration was high (log 8 = the concentration was ~10^8 CFU·mL^{-1}).

Another question arose during the experiments, namely whether during 1 h of contact all of the bacteria that were inoculated on the surface of the agar plate attached to the textile specimen. The answer to this is no, as suggested by the fact that after removing the specimen after 1 h and incubating the agar plate, the bacteria continued to grow over the area where the specimen had been (Figure 6).

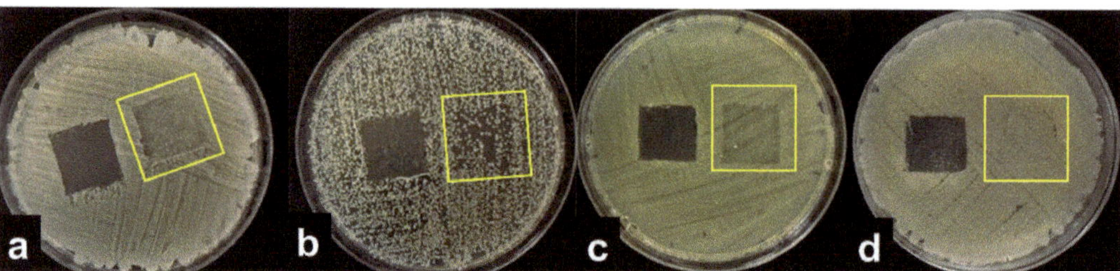

Figure 6. Results of the agar diffusion test in which the right specimen was removed after 1 h of contact and the left specimen was removed after 24 h of contact. (**a**,**b**)—experiments with wool and *S. aureus*; (**c**)—experiment with cotton and *A. baumannii*; (**d**)—experiment with compressed gauze and *B. cereus*.

Now that we have established that none of the materials exhibited bactericidal or bacteriostatic activity, we can compare whether bacteria attach with a different affinity on different materials.

Generally, there was no notable difference in the number of bacteria attached to either wool, cotton, or gauze after 1 h of contact. Few statistically significant differences are marked in Figure 7, perhaps indicating a lesser affinity for bacterial adsorption by cotton when compared to wool or gauze. However, in any case, neither material has shown to clearly have a higher capacity for bacterial adsorption during 1 h of contact. On the contrary, after 24 h of contact, gauze had a statistically significantly higher number of bacteria when compared to wool or cotton in almost every experiment (Figure 8). The only exception is the experiment with *A. baumannii* ATCC19606 strain (Figure 8). Wool had a higher number of bacteria when compared to cotton in experiments with *K. pneumoniae*, *B. cereus*, and *A. baumannii* strains ATCC19606 and HI2 (Figure 8). By comparing the three materials, it would seem that all the materials initially adsorb the same amount of bacterial cells. However, during the 24 h of incubation, the bacterial growth is most intensive on the gauze than on cotton or wool. Yet, gauze was shown to have the strongest growth under the samples in the agar tests.

From all the observations, the possible explanation of antibacterial activity as interpreted by the agar diffusion method could be the following, when placed on the agar surface, the bacteria attach to the textiles and continue their growth, both on the textile fibres and the agar surface under the specimen. After 24 h of incubation, when textile specimens are lifted off the agar surface, the bacteria remain attached to the textile fibres, probably developing a biofilm. In the case of the wool fabric, all of the bacteria are firmly attached to the material, and almost none remain on the agar surface, while in the case of the gauze, many bacteria remain on the agar surface below the sample. It would seem that different structures and compositions of materials have different affinities for the development of bacterial biofilm. In addition, the affinity for biofilm development did not seem to depend on the bacterial species. The formation of bacterial biofilms on textile fibers is a well-known phenomenon. The affinity for biofilm development was linked to the textile's hydrophobic and hygroscopic properties, surrounding environment, and bacterial species [20,21].

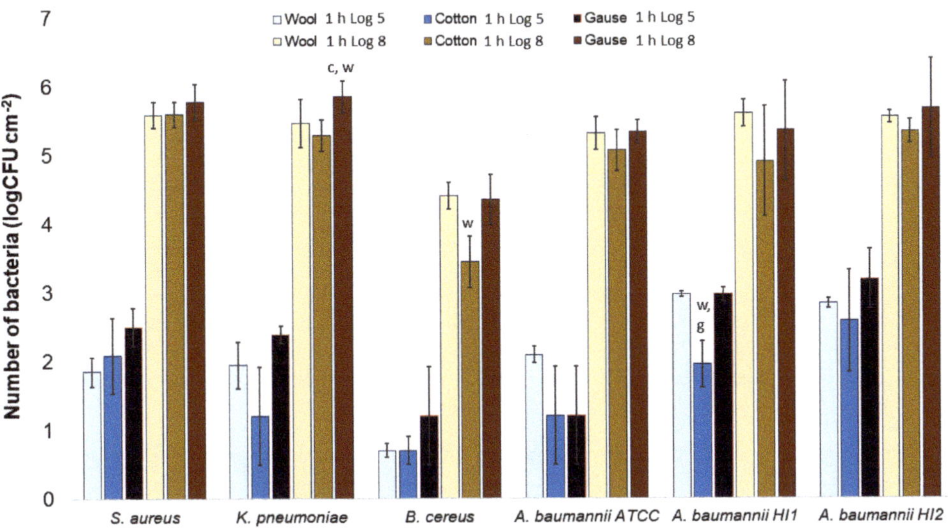

Figure 7. Numbers of viable bacteria that were adsorbed onto different specimens during the 1 h of contact in the agar diffusion test. Two experimental setups, one in which the inoculum concentration was low (log 5 = the concentration was ~10^5 CFU·mL^{-1}) and the other in which the inoculum concentration was high (log 8 = the concentration was ~10^8 CFU·mL^{-1}). w—statistically significantly different compared to wool; c—statistically significantly different compared to cotton; g—statistically significantly different compared to compressed gauze.

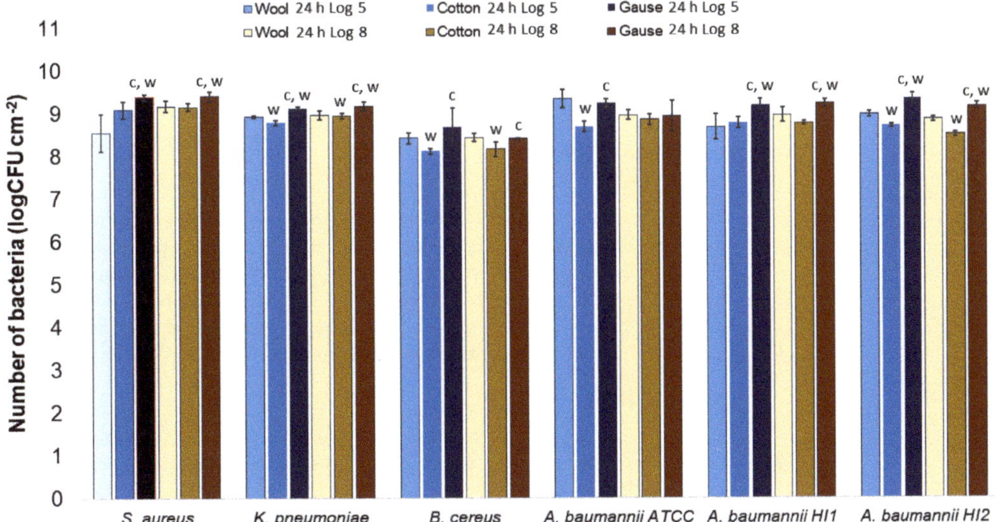

Figure 8. Numbers of viable bacteria that were adsorbed onto different specimens during the 24 h of contact in the agar diffusion test. Two experimental setups, one in which the inoculum concentration was low (log 5 = the concentration was ~10^5 CFU mL^{-1}) and the other in which the inoculum concentration was high (log 8 = the concentration was ~10^8 CFU mL^{-1}). w—statistically significantly different compared to wool; c—statistically significantly different compared to cotton; g—statistically significantly different compared to compressed gauze.

2.4. X-ray Microtomography

In our experiments, the antibacterial effect of the tested textiles resulted from the strong attachment of bacterial cells on the samples, especially the wool fabric. Therefore, we examined the surface topography of the tested materials by using X-ray Microtomography (XT). This method allows visualization of textile surfaces that were in contact with the agar surface on the mesoscale and not just on the microscale (i.e., few fibres) as with electron microscopy. In addition, the wool and cotton fabrics had a similar topography (Figure 9), while compress gauze differed significantly (Figure 9). The images could explain the apparent heavy growth under the gauze samples; the gauze material had a higher porosity than the other samples. Only a small part of the fibres are in direct contact with the agar surface, while wool and cotton were in contact with agar with most of their surface. The bacteria probably developed a strong biofilm only on the fibres in direct contact with the agar and thus were removed from the surface in the tests with wool and cotton and remained on the agar surface in the tests with the compress gauze leaving significant biomass under the gauze samples.

Figure 9. A 3D reconstruction of the surface of textile specimens used in the experiments, obtained by X-ray Microtomography.

3. Materials and Methods

3.1. Bacterial Strains

The experiments were performed using the *Staphylococcus aureus* (ATCC 25 923), *Klebsiella pneumoniae* (ATCC 11296), *Bacillus cereus* (LBK 4080), and *Acinetobacter baumanii* ATCC 19606 strain along with two multiple drug resistant hospital isolates, marked as HI1 and HI2. All the bacteria were kept cryo-stored using the MicrobankTM system (Pro-Lab diagnostics, Richmond Hill, ON, Canada). The bacteria were grown on Tryptic soy agar (TSA) plates (37 °C/24 h) prior to the start of the experiments.

3.2. Textile Materials

To test the antibacterial properties of untreated textiles, one sample of scoured wool fabric and two cotton samples were used in the experiment. The plain weave woolen fabric of linear density (for warp/weft) 15.5 × 2/25 tex, count (warp/weft) of 21/24 cm^{-1}, and mass per unit area of 140 g/m^2, is labeled as (W) and was industrially prepared (washed, decatised, sheared and dried) and supplied by Varteks Ltd. (Varazdin, Croatia). For comparison, two samples of cotton material of different degrees of finishing were also tested. The first cotton sample was a standard laboratory white coat (C) in twill with an embroidery count (warp/weft) of 18/17 cm^{-1} and mass per unit area of 220 g/m^2, manufactured by Marija Ltd. (Zagreb, Croatia). The second one was a standard sterile compressed gauze (G) in canvas embroidery, with a count (warp/weft) of 11/8 cm^{-1} and mass per unit area of 40 g/m^2, manufactured by Lianyungang Ruikang Sanitary Dressing Company Ltd. (Lianyungang, China). For a positive control, a textile with known antibacterial properties, Aquacell® wound dressing that is incorporated with ionic silver (Convatec Inc., Berkshire, UK). The textiles were cut into 20 × 20 mm squares and used in the experiment without prior sterilization, as noted in ISO 20645:2004 standard method. However, disinfected nitrile hand gloves were used during the cutting to minimize contamination of samples from the skin bacteria.

3.3. Antibacterial Testing

The procedure of antibacterial testing (Figure 1) was based on a method described in ISO 20645:2004—"Textile fabrics—Determination of antibacterial activity—Agar diffusion plate test" (reference ISO 2004). Suspensions of concentration 10^5 and 10^8 Colony Forming Units (CFU) per mL of sterile 0.3% saline were made for each tested bacterium. Such solutions were made by dispersing bacterial biomass in sterile saline up to 0.5 McFarland units (corresponding to ~10^8 cells per mL). Next, the solution was serially diluted to obtain 10^5 CFU mL^{-1} suspension. The CFU's were checked by plating prior to each experimental batch. To grow the "bacterial lawn", bacterial suspension was spread across the Tryptic soy agar (Biolife, Italy) plate using a sterile cotton swab. After plate inoculation, two textile specimens (20 × 20 mm) were placed on the agar surface using sterile tweezers. The agar plates were then incubated for 24 h at 37 °C. The results were interpreted according to ISO 20645:2004 standard by examining colony growth under the textile sample, visually and under the microscope (Olympus Japan, CX21) at 40× magnification (Figure 1).

3.4. Determining the Number of Bacteria Attached to Textile Samples

To determine if and in what amount the bacteria remain attached to textiles during the antibacterial testing (as previously described), the samples were gently removed (Figure 1) and immersed in 20 mL of sterile saline (in 50 mL Falcon-type tubes). The tubes were shaken-out on a vortex shaker according to ISO 20743:2013—"Textiles—Determination of the antibacterial activity of textile products", for 5 × 5 s cycles. Shaking detaches the bacteria from the fabric, and the cells remain free-floating in the saline suspension. A total of 1 mL of suspension was serially diluted up to 10^{-7}, and 0.1 mL was inoculated on TSA plates. After the incubation (24 h/37 °C) the grown colonies were counted, and the bacterial numbers were reported as CFU per cm^2 of textile material. As a control, clean textile specimens were used, meaning they were not previously incubated on agar plates. Several bacterial colonies usually grew as the specimens were not sterile, but total counts were less than 10 CFU·mL^{-1}.

3.5. X-ray Microtomography

X-ray microtomography was performed to visualize the inner structure of the three types of textiles at the micron scale. The 3D images of the textiles were obtained on a laboratory tomograph manufactured by RX Solutions (Annecy, France) equipped with a Hamamatsu X-ray source (Hamamatsu City, Japan) and a Varian flat panel detector (Varian Medical Systems, Salt Lake City, UT, USA). Each sample was irradiated with an X-ray beam (generated with a 100 kV 100 µA electron beam on a tungsten target) for 2400 angular projections equally spaced over 360°. The 2D radiographs were converted into a 3D dataset using a filtered back projection algorithm, and 3D views were obtained using the 3D viewer of Fiji software (v 1.52f). The chosen pixel size was set to 10 µm. This pixel size was chosen as it allows the visualization of the fibers and the fiber bundle that constitute the textile and a representative volume of the textile as many periods of the structure can be seen.

3.6. Statistical Analysis

All the experiments were undertaken in triplicate. The growth under the textile specimen was determined qualitatively by visual inspection. The numbers of bacteria attached to textile samples were quantitatively compared and analyzed using Statistica® software (StatSoft, Tulsa, OK, USA). Ordinary Student's t-test was used, and statistical significance was set at $p < 0.05$.

4. Conclusions

The wool fabric showed antibacterial efficacy towards several bacterial species if interpreted according to the agar diffusion test as "no growth" under the textile sample. On the other hand, experiments monitoring the number of viable cells after 24 h of bacteria/textile contact showed that neither the wool sample nor two different cotton samples exhibited any bactericidal or bacteriostatic activity in terms of inactivating or killing bacterial cells. Instead, bacterial cells readily multiplied on the textile samples during 24 h of incubation on nutrient agar plates. The explanation would thus be that bacteria strongly adsorb to wool while actively multiplying, developing a firmly attached biofilm. When the wool sample was lifted off, the bacterial biofilm remained attached to wool fibers, removing the biomass from the surface of the nutrient agar, resulting in a clear "no growth" zone under the sample.

Since similar observations were present in experiments with cotton but not with compress gauze, it would seem that the surface topography and structure of the textile plays an important role in the antibacterial efficacy of the textiles which are unmodified with some antibacterial agent. The results indicate that certain textiles, in our case woven wool fabric, could exhibit antibacterial properties by removing excess bacteria that grow on the textile/skin interface when taken off the body.

A deeper analysis of biofilm formed on the textile fibers, visualized by scanning electronic microscopy, and quantified as it develops, would also be desirable. Along with further tests with different non-modified textile materials and the same materials of differing structure and porosity, the question of why some textiles seem to show antibacterial efficacy without showing any bactericidal activity could be resolved.

Author Contributions: T.I. and S.E.R., conceptualization, experimental design, writing original draft preparation, review and editing.; A.R., methodology and experimental setup.; S.R.d.R., X-ray Microtomography analysis and editing.; Z.S., project administration and funding acquisition. All authors have read and agreed to the published version of the manuscript.

Funding: This work has been fully supported by the Croatian Science Foundation under the project (IP-2016-06-5278). The X-ray Microtomography imaging was enabled with the support of "COGITO" Hubert-Curien Partnership (Campus France) program no. 42815QE. The 3SR is part of LabEx Tec 21—ANR-11-LABX-0030 and of Institut Carnot PolyNat (ANR16-CARN-0025).

Institutional Review Board Statement: Not applicable.

Informed Consent Statement: Not applicable.

Data Availability Statement: Not applicable.

Conflicts of Interest: The authors declare no conflict of interest.

Sample Availability: Samples of the compounds are available from the authors.

References

1. Caven, B.; Redl, B.; Bechtold, T. An investigation into the possible antibacterial properties of wool fibers. *Text. Res. J.* **2019**, *89*, 510–516. [CrossRef]
2. Holcombe, B. Wool performance apparel for sport. In *Advances in Wool Technology*; Johnson, N.A.G., Russell, I.M., Eds.; Woodhead Publishing Limited: Cambridge, UK, 2009; pp. 265–283.
3. Johnson, N.A.G.; Wood, E.J.; Ingham, P.E.; McNeil, S.J.; McFarlane, I.D. Wool as a Technical Fibre. *J. Text. Inst.* **2003**, *94*, 26–41. [CrossRef]
4. Yu, D.; Tian, W.; Sun, B.; Li, Y.; Wang, W.; Tian, W. Preparation of silver-plated wool fabric with antibacterial and anti-mould properties. *Mater. Lett.* **2015**, *151*, 1–4. [CrossRef]
5. Dickerson, M.B.; Sierra, A.A.; Bedford, N.M.; Lyon, W.J.; Gruner, W.E.; Mirau, P.A.; Naik, R.R. Keratin-based antimicrobial textiles, films, and nanofibers. *J. Mater. Chem. B* **2013**, *1*, 5505–5514. [CrossRef] [PubMed]
6. Peran, J.; Ercegovic Razic, S.; Sutlovic, A.; Ivankovic, T.; Glogar, M.I. Oxygen plasma pretreatment improves dyeing and antimicrobial properties of wool fabric dyed with natural extract from pomegranate peel. *Color. Technol.* **2020**, *136*, 177–187. [CrossRef]

7.	McNeil, S. *The Removal of Indoor Air Comtaminants by Wool Textiles*; Technical Bulletin; AgResearch: Christchurch, New Zealand, 2015. [CrossRef]
8.	Yilmaz, E.; Celik, P.; Korlu, A.; Yapar, S. Determination of the odour adsorption behaviour of wool. *Text. Leather Rev.* **2020**, *3*, 30–39. [CrossRef]
9.	Causer, S.M.; McMillan, R.C.; Bryson, W.G. The role of wool carpets and furnishings in reducing indoor air pollution. In Proceedings of the 9th International Wool Textile Research Conference, Biella, Italy, 28 June–5 July 1995; pp. 155–161.
10.	Pinho, E.; Magalhães, L.; Henriques, M.; Oliveira, R. Antimicrobial activity assessment of textiles: Standard methods comparison. *Ann. Microbiol.* **2011**, *61*, 493–498. [CrossRef]
11.	Benesovsky, P. Determination of antibacterial activity of textiles—Methods, results and their interpretation. In Proceedings of the ICAMS 2010—3rd International Conference on Advanced Materials and Systems, Bucharest, Romania, 16–18 September 2010.
12.	*AATCC 147:2004*; Antimicrobial Activity Assessment of Textile Materials: Parallel Streak Method. American Association of Textile Chemists and Colorists: Research Triangle Park, NC, USA, 2004.
13.	Liu, X.; Lin, T.; Peng, B.; Wang, X. Antibacterial activity of capsaicin-coated wool fabric. *Text. Res. J.* **2012**, *82*, 584–590. [CrossRef]
14.	Gomes, A.P.; Mano, J.F.; Queiroz, J.A.; Gouveia, I.C. Layer-by-layer deposition of antimicrobial polymers on cellulosic fibers: A new strategy to develop bioactive textiles. *Polym. Adv. Technol.* **2013**, *24*, 1005–1010. [CrossRef]
15.	Liu, H.; Lee, Y.Y.; Norsten, T.B.; Chong, K. In situ formation of anti-bacterial silver nanoparticles on cotton textiles. *J. Ind. Text.* **2014**, *44*, 198–210. [CrossRef]
16.	Li, Z.; Tang, H.; Yuan, W.; Song, W.; Niu, Y.; Yan, L.; Yu, M.; Dai, M.; Feng, S.; Wang, M.; et al. Ag nanoparticle–ZnO nanowire hybrid nanostructures as enhanced and robust antimicrobial textiles via a green chemical approach. *Nanotechnology* **2014**, *25*, 145702. [CrossRef] [PubMed]
17.	Haase, H.; Jordan, L.; Keitel, L.; Keil, C.; Mahltig, B. Comparison of methods for determining the effectiveness of antibacterial functionalized textiles. *PLoS ONE* **2017**, *12*, e0188304. [CrossRef] [PubMed]
18.	Pollini, M.; Paladini, F.; Licciulli, A.; Maffezzoli, A.; Nicolais, L.; Sannino, A. Silver-coated wool yarns with durable antibacterial properties. *J. Appl. Polym. Sci.* **2012**, *125*, 2239–2244. [CrossRef]
19.	Abdel-Fattah, S.H.; El-Khatib, E.M. Wool fabrics with antibacterial properties. *RJTA* **2012**, *16*, 42–48. [CrossRef]
20.	Moellebjerg, A.; Meyer, R. The bacterial lifecycle in cotton and polyester textiles. In Proceedings of the Biofilms 9 Conference, Karlsruhe, Germany, 29 September–1 October 2020. [CrossRef]
21.	Sanders, D.; Grunden, A.; Dunn, R.R. A review of clothing microbiology: The history of clothing and the role of microbes in textiles. *Biol. Lett.* **2021**, *17*, 20200700. [CrossRef] [PubMed]

Article

Influence of Cotton Pre-Treatment on Dyeing with Onion and Pomegranate Peel Extracts

Lea Botteri, Anja Miljković and Martinia Ira Glogar *

Faculty of Textile Technology, University of Zagreb, 10000 Zagreb, Croatia; lea.botteri@ttf.hr (L.B.); amiljkovi@ttf.hr (A.M.)
* Correspondence: martinia.glogar@ttf.hr; Tel.: +385-14877364

Abstract: In this paper the possibility of applying natural dyes on cellulose fibres were researched with respect to the impact of cotton material pre-treatment (scouring, chemical bleaching, mercerization and mordanting), using renewable sources of natural dyes (waste as a source). As mordants, metal salts of copper, aluminium and ferrum were used, and the influence on colour change as well as on fastness properties were analysed. The natural dyes were extracted from onion peel (*Allium cepa* L.) and pomegranate peel (*Punica granatum* L.). In spectrophotometric analysis performed of the plant extracts, the onion extract has peaks at 400 and 500 nm, resulting in red-orange colourations and the pomegranate extract shows a maximum at 400 nm, i.e., in the yellow region, which is characteristic of punicalin. Results show significant influence of cotton pre-treatments on colour appearance and fastness properties, caused by pre-treatments affecting the properties and structure of the cotton itself. The positive effect of mercerization on dye absorption and bonding is confirmed. For wash and light fastness properties, more satisfactory results have been obtained for yarns dyed with pomegranate peel natural dye, and the key importance of mordants for fastness properties has been confirmed.

Keywords: cotton yarn; scouring; chemical bleaching; mercerization; natural dyes

Citation: Botteri, L.; Miljković, A.; Glogar, M.I. Influence of Cotton Pre-Treatment on Dyeing with Onion and Pomegranate Peel Extracts. *Molecules* **2022**, *27*, 4547. https://doi.org/10.3390/molecules27144547

Academic Editor: Baljinder Kandola

Received: 19 May 2022
Accepted: 5 July 2022
Published: 16 July 2022

Publisher's Note: MDPI stays neutral with regard to jurisdictional claims in published maps and institutional affiliations.

1. Introduction

Recent decades have witnessed scientific research efforts aimed at revitalizing and commercializing the use of natural dyes in textile dyeing. There are many positive aspects of natural dyes, which justify this work and effort: from the positive colour characteristics of which the colour hues achieved by natural dyes are always in a harmonious and balanced relationship that positively affects the user, to their proven antibacterial properties, protective properties form harmful UV radiation and environmental sustainability and non-toxicity [1–5]. However, the application of natural dyes in commercial production has certain obstacles. Natural dyes many times lack in uniformity and reproducibility of colour. In addition, the problem is their availability in bulk quantity. There is no sufficient information on standardization of natural dyes application methods, as well as often-questionable colourfastness properties [6]. The application of natural dyes on cotton attracts special attention due to the popularity of cotton fibre, but this is where most of the problems are encountered, due to the natural affinity of most natural dyes for not cellulose but protein fibres. Ferreira conducted a comprehensive study of flavonoid dyes, among which emphasizes research into the chemical structure of onion peel dyes. The results provided important structural information on previously unidentified flavonoids in natural yellow dye extracts. The study contributes greatly to the understanding of the responsibility of individual compounds contained in the dye structure of onion peel, to achieve a certain colour hue, but also for the fastness properties, mostly light fastness [7]. Ferreira et al. also published a comprehensive study of the chemical constituents of the main classes of natural dyes, pointing out the importance of the possibilities provided by modern experiment methodology in establishing chemical correlations with historical records, and the importance of understanding the sensitivity of individual compounds to photo-degradation, thus allowing a more precise definition of the conditions of storage of historical textiles [8]. They also researched the identification and photochemical degradation of flavonoid dyes using photo-oxidation products of quercetin and morin as a marker for the characterization of natural yellow dyes in ancient textiles [9].

Papers dealing with topics similar to this paper confirms the extensive research work invested in studying the mechanisms of dyeing cellulosic materials with natural dyes and their binding to cellulose, as well as the influence of mordant, not only on significant colour changes, but also in the context of their key role in achieving satisfactory colour fastness. Iqbal and Liaqat researched the possibility of applying the natural dye extracted from pomegranate peel on cotton fabric and achieved a satisfactory results of colour depth, intensity and wash fastness properties by using the oxalic acid as mordant [10]. The antibacterial, biocidal and deodorizing properties of the natural dye obtained from pomegranate peel and applied at cotton have been also confirmed [11,12]. Davulcu et al. [13] investigated the possibility of applying natural dyes of thyme and pomegranate peel on cellulose fabrics, analysing the influence of mordant on the colour characteristics as well as on washing, rubbing, perspiration and light fastness properties. Using potassium aluminium sulfate, copper (II) sulfate, iron (II) sulfate and tin (II) chloride as mordant, interesting conclusions were reached about minimal or even no effect of mordant on wash fastness of samples dyed with thyme dye. For pomegranate peel dyed samples, also the premordanting process did not enhance the washing, rubbing or perspiration fastness properties, which was satisfactory already on non-mordanted samples. As for the light fastness, they confirmed the key role of mordants in achieving the satisfactory results. They also confirm the antibacterial properties of the samples dyed without prior mordanting. The similar work has been conducted by Kulkarni et al. [14] with the proviso that they analysed the impact of pre-, during dyeing- and post-mordanting. Satyanarayana and Remesh Chandra [15] researched the possibility of applying pomegranate peel dyes on cotton fabrics with prior scouring of cotton fabric and mordanting with copper sulphate and ferrous sulphate. They obtained moderate rub, good wash and very good light fastness of dyed samples. Adeel et al. [16] focused on optimizing the dyeing process with natural dye of pomegranate peel by monitoring the influence of dyeing temperature and time and salt concentration as an additive in the dyeing bath. In their conclusion, they also confirmed that despite ongoing research, there are still many open areas where a solution needs to be found. Silva et al. [17] investigated the application of onion skin dye on cotton fabric with the possibility of using chitosan as the replacement for electrolyte in dyeing process. As the results they obtained acceptable levels of fastness and antimicrobial as well as finding that chitosan contributed to an increase in UPF value as the measure of protective properties against harmful UV exposure. Such research confirms great interest in the field of application of natural dyes on textiles as well as finding the possibility of returning to natural dyes, not only in localized production of small series but also in wider, even industrial production.

Regarding the application of natural dyes in modern production, the important aspect is the sustainability of dye sources. Cheap by-products from agriculture and forestry can be used to obtain natural plant dyes, e.g., wood bark from the timber industry, waste from the food and beverage industry such as pressed berries, distillation sludge and other residues. The source of natural dyes can also be, for example, wastewater from olive mills, which is a by-product of olive oil extraction and causes major environmental problems in Mediterranean countries. An intensive use of industrial wastes as renewable raw materials for the production of natural dyes would increase the economic use of waste materials, contribute to the protection of the environment and reduce the use of fossil fuels [18–24]. This approach to the recovery of waste as a source of raw materials fits perfectly into the zero-waste concept, which implies integrated recycling, material reuse, resource optimization, waste reduction and deconstruction.

In this paper, the complex issue of applying natural dyes on cellulosic materials is addressed from two aspects: first, the possibility of applying natural dyes on cellulose fibres with respect to their natural affinity for protein fibres and the impact of cotton material pre-treatment (scouring, chemical bleaching, mercerization and mordanting) and second, the possibility of using renewable sources of natural dyes (waste as a source). The plants used, onion and pomegranate, traditionally grow in Croatia. Extracts obtained from the waste of these plants were used in dyeing cotton yarn. Raw, scoured, chemically bleached, mercerized and mordanted cotton yarns were dyed with natural dyes obtained from onion peel (*Allium cepa* L.) and pomegranate peel (*Punica granatum* L.). Cotton yarns dyed with natural dyes has added value not only for environmental and economic reasons but also from a social and psychological aspect. This research contributes to the complex problem of dyeing cellulose fibres with natural dyes, but also to the circular economy.

2. Methods

The 100% raw cotton yarns®, Nm 16/2, Z-15 (Unitas, Zagreb, Croatia) were scoured (S), chemically bleached (CB) and mercerized (RM, RM4). After the pre-treatments, the cotton yarns were dyed with onion peel (*Allium cepa* L.) and pomegranate peel (*Punica granatum* L.). The labels and treatments are shown in Table 1.

Table 1. Labels and treatments.

Labels	Treatments
R	Raw cotton yarn
S	Scoured cotton yarn
CB	Chemical bleach cotton yarn
RSM	Raw slack mercerized yarn
RM4	Raw mercerized yarn with 4% of tenacity
*O**	Cotton yarn dyed with onion peel
*P**	Cotton yarn dyed with pomegranate peel
*Al	Cotton yarn mordanted with: $KAl(SO_4)_2 \cdot 12H_2O$
*Cu	Cotton yarn mordanted with: $CuSO_4 \cdot 5H_2O$
*Fe	Cotton yarn mordanted with $FeSO_4 \cdot 7H_2O$

*—Raw cotton yarn (R), Scoured cotton yarn (S), Chemical bleach cotton yarn (CB), Raw slack mercerized yarn (RSM), Raw mercerized yarn with 4% of tenacity (RM4) dyed with onion peel (O) or pomegranate peel (P). **—Raw cotton yarn (R), Scoured cotton yarn (S), Chemical bleach cotton yarn (CB), Raw slack mercerized yarn (RSM), Raw mercerized yarn with 4% of tenacity (RM4) dyed with onion peel (O) or pomegranate peel (P) without or with mordanted ($KAl(SO_4)_2 \cdot 12H_2O$, $CuSO_4 \cdot 5H_2O$, $FeSO_4 \cdot 7H_2O$).

2.1. Pre-Treatments of Cotton Yarns

Scouring was performed in sodium hydroxide solution with the addition of wetting. The natural pigments remaining after scouring were removed by chemical bleaching with hydrogen peroxide in an alkaline medium with the addition of organic and inorganic stabilizers, sequestering agent and antifoaming agent. Slack mercerization was performed in a laboratory beaker containing 24% sodium hydroxide solution with the addition of a wetting agent. Cotton yarns for mercerization with tenacity were placed in the frame of the mercerizing machine and immersed in 24% sodium hydroxide solution with the addition of a wetting agent. The cotton yarns were twisted in both directions at 4% of tenacity. The pre-treatment procedures are listed in Table 2.

Table 2. Pre-treatment procedures.

Pre-Treatment	Scouring	Chemical Bleaching	Mercerization (Slack/with Tenacity 4%)
Batch Composition	• NaOH (3%) • Felosan (CHT Group): 2g/L	• H_2O_2 (35%): 25 mL/L • NaOH: 4 g/L • Tinoclarit CBB (Ciba, Swiss): 5 mL/L • Water glass (mixture of Na_2SiO_3, $Na_2Si_2O_5$): 15 mL/L • Heptol ESW (CHT Group): 10 mL/L • Fumexol DF (Ciba, Swiss): 0.5 mL/L	• NaOH (24%) • Subitol MLF (Bezema): 8 g/L
Methods	Process of exhaustion	Process of exhaustion	• Slack mercerization: laboratory beaker • Mercerization with tenacity 4%: mercerizing machine
Proces Parameters	T = 100 °C t = 30 min	T = 100 °C t = 30 min	T = 16 °C t = 120 s
Bath Ratio	1:20	1:20	/
After Pretreatmant	• Rinsing (hot soft water, medium hot water and cold soft water)	• Rinsing (hot soft water, medium hot water and cold soft water) • Neutralization (1% CH_3COOH) • Rinsing (soft water until neutral)	• Rinsing (hot soft water, medium hot water and cold soft water) • Neutralization (1% CH_3COOH) • Rinsing (soft water until neutral)

2.2. Dye Extraction

Onion peel and pomegranate peel were used for dye extraction (Figure 1). Onion peel (*Allium cepa* L.) is rich in flavonoids, the most abundant of which is quercetin (2-(3,4-dihydroxyphenyl)-3,5,7-trihydroxy-4H-chromen-4-one), followed by quercetin glucoside and kaemfoerol. Due to their chemical structure, flavonoids are the most abundant plant dyes of yellow hue. Pomegranate peel (pomegranate, *Punica granatum* L.) contains 28% tannins. The hydrolysate may be gallic acid, ellagic acid and flavogalol. The tannins are punicalagin (2,3-(S)-hexahydroxydiphenoyl-4,6-(S,S)-galagyl-D-glucose; α-punicalagin; β-punicalagin) and puniclin (4,6-(S,S)-galagyl-D-glucose), which are responsible for the yellow colour [25,26].

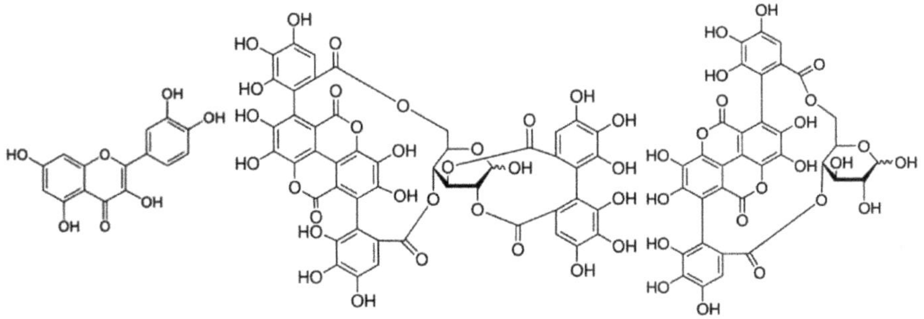

Figure 1. Chemical structures: (**a**) quercetin in onion peel and tannin derivatives in pomegranate peel; (**b**) punicagalin; (**c**) punicalin.

Dye extraction was performed in soft water containing 100 g/L onion peel and pomegranate peel. The extraction was performed in a 1:40 bath ratio (considering the mass of the plant) at 100 °C for 60 min. The bath was then allowed to cool for 12 h and the extract was decanted.

Spectrophotometric analysis of the plant extracts was performed on a Cary 50 Solascreen, Varian absorption spectrophotometer in the ultraviolet (250–400 nm) and visible parts of the spectrum (400–700 nm).

2.3. Mordanting of the Textile Material with Metal Salts

Onion peel and pomegranate peel belong to the group of acid mordant dyes due to their dyeing properties and bind to fabric by forming a complex with metal salts, called mordants. For this reason, some samples were treated with potassium aluminium sulphate dodecahydrate $KAl(SO_4)_2·12H_2O$, copper (II) sulphate pentahydrate $CuSO_4·5H_2O$ and ferrous (II) sulphate heptahydrate $FeSO_4·7H_2O$; (Kemika, Zagreb, Croatia) before the dyeing process [27].

Pre-treatment of yarns with metal salts was performed with 5% mordants (based on the mass of the material) in a bath ratio of 1:30 in a Polycolor Mathis apparatus at 50 °C for 30 min. After pre-treatment with metal salts, the cotton yarns were rinsed with cold water.

2.4. Dyeing with Natural Dyes

Dyeing with natural dyes using onion peel and pomegranate peel of pre-treated cotton yarns was performed with a bath ratio of 1:30 in a Polycolor Mathis apparatus at 60 °C for 60 min. After dyeing, the cotton yarns were rinsed with cold water. Since these natural dyes are from the group of acid-mordant dyes, dyeing was performed at a pH of 4 adjusted with 20% acetic acid (Kemika, Zagreb).

2.5. Colour Analysis in the CIEL*a*b* System

Colour characteristics were measured using a remission spectrophotometer, Datacolor 850, measuring geometry d/8°, D65, measuring aperture of 9 mm. The whiteness (W_{CIE}) of undyed cotton yarns were performed according to ISO 105-J02:1997 Textiles—Tests for colour fastness—Part J02: Instrumental assessment of relative whiteness. The coordinates used to determine colour values are "L*" for lightness, "a*" for redness (positive value) and greenness (negative value), "b*" for yellowness (positive value) and blueness (negative value), "C*" for chroma and "h" for hue angle in the range of 0° to 360° of undyed and dyed cotton yarns were determined according to ISO 105-J01:1997 Textiles—Tests for colour fastness—Part J01: General principles for measurement of surface colour.

All results were measured on samples by repeating the measurement procedure at random locations on the samples. Thus, the colour measurements were made using the Datacolor Tools computer program and "Measuring until tolerance" command, which means that at least 10 measurements must be made, and the results are accepted only if the total colour difference between each measurement is less than 0.1 (dE* < 0.1).

2.6. Wash Fastness

Wash fastness of cotton yarns were tested in a laboratory apparatus for wet and dyeing processes Polycolor, Mathis. The test was performed according to standard ISO 105-C06:2010 (A2S) Textiles—Tests for colour fastness—Part C06: Colour fastness to domestic and commercial laundering, using 2 g/L of standard detergent (James Heal ECE A, without optical brighteners and without phosphates), with a bath ratio of 1:20, temperature of 40 °C, time of 30 min. The results of wash fastness are given as numerical values of total colour difference calculated according to CIE76 equation, as well as in grey scale grades, obtained by comparing unwashed samples with samples that were washed after the 1st and 5th washing cycle.

2.7. Lightfastness

Lightfastness testing of the samples were performed on Xenotest 440 ((SDL Atlas, Rock Hill, SC, USA). Xenotest 440 is used for laboratory simulation of external weather influences on the stability and durability of textile and other materials. Analysis was evaluated according to the modified ISO 105-B02 and 13 B04 test methods using Xenotest 440. Test conditions simulated in this research were: Total light time: 41:10 h, Radiant exposure: 6226 kJ/m^2, Irradiance control: 300–400 nm, Filter system: B04, E: 42 W/m^2 (± 2 W/m^2), CHT: 32 °C (± 3 °C), BST: 47 °C (± 8 °C), RH: 40% (± 8%), no spray, fan speed: 2000 rpm. Using the same equation as for the wash fastness, the lightfastness properties was also evaluated by calculating total colour difference values (DE), as well as in blue scale grades, obtained by comparing samples before and after exposure.

3. Results and Discussion

In this paper the influence of pre-treatment of cotton material on dyeing with natural dyes is presented. Pre-treatments of scouring, chemical bleaching and mercerization were performed. After the process of scouring, a part of the samples was chemically bleached and a part was left only scoured, while the mercerization process was carried out directly on the raw samples without prior scouring and bleaching.

In the first step, the spectrophotometric analyses of whiteness (W), yellowness (YI) and lightness (L*) have been performed on undyed samples, regarding the different pre-treatments. The objective results of whiteness, yellowness and lightness of undyed treated samples are shown in Table 1, while the photographs of samples are given in Table 3.

Table 3. Whiteness, yellowness index and spectral characteristics of pre-treatment and mordanted yarns.

Labels	W_{CIE}	YI	L*	a*	b*	C*	h°
R	−17.00	34.52	84.42	2.92	16.48	16.74	79.95
R_Cu	−3.10	25.18	82.68	−0.40	12.64	12.65	91.80
R_Al	−3.00	29.03	84.60	2.28	13.74	13.93	80.58
R_Fe	−19.20	32.49	81.03	2.28	15.08	15.25	81.42
S	40.10	14.19	88.03	1.17	6.64	6.74	79.98
S_Cu	39.20	13.59	88.38	−0.28	6.97	6.98	92.31
S_Al	47.10	12.92	90.16	0.85	6.25	6.30	82.25
S_Fe	−0.50	28.88	85.54	2.42	13.73	13.94	80.01
CB	79.70	1.75	93.49	−0.20	0.96	0.98	101.94
CB_Cu	78.10	−0.86	90.98	−1.25	0.05	1.25	177.57
CB_Al	85.40	1.56	95.71	−0.18	0.87	0.89	101.97
CB_Fe	21.30	23.71	90.92	1.12	13.13	12.18	84.73
RSM	5.60	24.31	83.44	1.52	11.38	11.48	82.41
RSM _Cu	−1.50	21.24	79.86	−1.95	10.96	11.13	100.8
RSM_Al	14.00	22.51	85.99	0.97	10.96	11.01	84.96
RSM_Fe	−8.30	29.68	83.21	1.95	14.04	14.18	82.09
RM4	1.40	24.66	81.85	1.38	11.44	11.52	83.13
RM4_Cu	11.00	18.77	82.37	−1.48	9.73	9.84	98.62
RM4_Al	6.20	23.67	83.20	1.23	11.14	11.21	83.69
RM4_Fe	−35.50	38.54	79.92	3.37	17.64	17.96	79.17

In addition to cellulose, cotton fibre contains impurities that give cotton its hydrophobicity and thus prevent satisfactory treatments of cotton material. The scouring has been performed in order to remove the hydrophobic impurities from the primary wall (e.g., pectin, proteins, and organic acids) and cuticle (waxes and fats) from cotton fibre [25,28–30]. The spectrophotometric measurement of whiteness of raw material showed rather low value (W_{CIE} = −17.00), which is to be expected since raw cotton has impurities that give it a yellowish colour. Scouring removes all impurities except pigments, and whiteness increases (W_{CIE} = 40.10), while the highest whiteness is obtained with chemical bleaching (W_{CIE} = 79.70), causing the natural pigments to be removed, resulting in a significant increase in whiteness. In this way, the results of dyeing become accurate. Due to the protoplasmic residues of the protein and flavone pigments of the cotton flowers, it has its natural greyish colour. By chemical bleaching, the cotton obtains a permanently white surface suitable for dyeing [26]. The yellowness index is inversely related to the whiteness, i.e., the higher the whiteness, the lower the yellowness index. The raw cotton yarn has the value of yellowness index (YI = 34.52) and the chemically bleached has the (YI = 1.75).

The raw cotton samples were also mercerized. Mercerization of cotton is performed in 20–30% NaOH solution with or without tension that achieves various effects such as increased moisture absorption, lustre and dyeability. The cotton fibre's longitudinal view is converted from a ribbon shape to a straight shape, and the natural twist of cotton fibres disappears. Moreover, part of the crystalline region of the cotton fibres is converted into the amorphous region, so that the dyeability and chemical reactivity are improved. The mercerized cellulose has no chemical changes compared to the original cellulose, but the crystal structure is converted from cellulose I to cellulose II [27,31–33].

The samples in this experimental work have been mercerized by both methods, in slack (without the tension) and with tension. The results show the slight increase in whiteness compared to raw samples, although the mercerized samples were not scoured and bleached previously. This is due to the action of the sodium hydroxide, which removes the impurities that give the cotton a yellowish hue. The whiteness of sample mercerized in slack mercerization is W_{CIE} RSM = 5.60 and mercerization with tenacity is W_{CIE} RM4 = 1.40.

By processing pre-treated yarns with metal salts of copper, aluminium and ferrum (mordants) the whiteness and yellowness index also change. Mordants in addition to improving the binding of dyes to fibre, also have a major role in changing the whiteness, yellowness index and hue of undyed pre-treated yarns. Influence of pre-treatment with copper and aluminium, greatly affects the increase in whiteness compared to pre-treated yarns. All cotton yarns give a yellowish shade, except for chemically bleached copper-treated yarns. The hue of CB_Cu is in the green area ($h° = 177.57$) due to the natural greenish shade of $CuSO_4 \cdot 5H_2O$ solution. When cotton yarn was treated with ferrum mordant, the yellowness index increases relative to all pre-treatment yarns except raw yarn. The largest difference in yellowness index can be seen on RM4_Fe, which is YI = 38.54. Accordingly, the RM4_Fe yarn has the lowest lightness ($L^* = 79.92$). Due to the removed pigments in chemically bleached yarns, the lightness is the highest and ranges from 93.49 to 90.92 (Tables 3 and 4).

Table 4. Pre-treated and mordanted cotton yarns.

Treatments	Mordants			
	-	Al	Cu	Fe
R				
S				
CB				
RSM				
RM4				

In the next step, the extraction of natural dyes from onion and pomegranate peel has been performed. As mentioned in the Introduction, one of the aims of this work was to investigate the possibility of using easily accessible natural sources (biowaste from food sources): pomegranate peel and onion peel for dyeing cotton yarn. The use of biowaste to obtain natural dyes is in line with one of the essential postulates of the circular economy, and that is waste minimization. It solves some of the basic problems for the possible commercial application of natural plant dyes: the need for plant sources, availability of raw materials, standardization of dyeing recipes, environmental and economic sustainability, i.e., the realization of the idea of zero greenhouse gas emissions. The idea of zero emissions is based on the idea that each biological waste can be a raw material for another production, i.e., that one industry can always consume the waste of another industry. Inexpensive by-products from agriculture and forestry can be used to obtain natural plant dyes, such as wood bark from the wood industry, industrial food and beverage waste such as pressed berries, distillation sludge and other residual by-products. Intensive use of industrial waste as renewable raw materials for the production of natural dyes would increase the economical use of waste material, contribute to the preservation of the environment and reduce the use of fossil fuels [34].

After the extraction, the absorption spectra of the extracts were measured spectrophoto-metrically. Figure 2 shows the absorption spectra of the extracts of onion and pomegranate peel in the ultraviolet and visible parts of the spectrum. In the visible part of the spectrum, the onion extract was confirmed by peaks at 250, 288 and 330 nm [35,36].

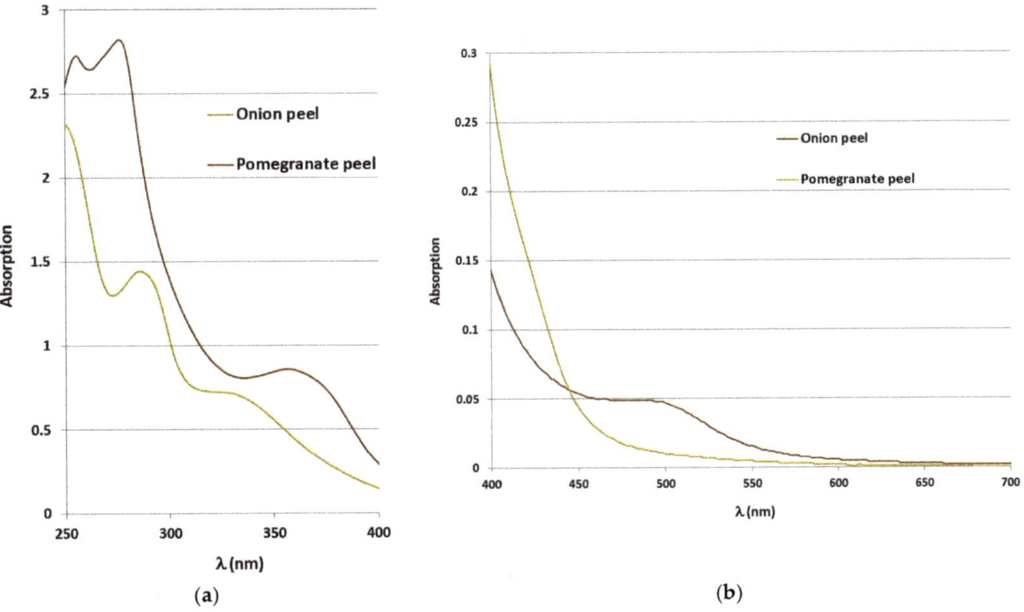

Figure 2. Absorption spectra of onion and pomegranate extracts in the (**a**) ultraviolet part and (**b**) visible part of the spectrum.

The pomegranate extract shows a maximum at 400 nm, i.e., in the yellow region, which is characteristic of punicalin. Punicalin in the aqueous extract of pomegranate was also confirmed by analysis in the ultraviolet part of the spectrum with peaks at 256, 278 and 360 nm [35,36].

Pomegranate peel (*Punica granatum* L.) contains 28% tannins. Pomegranate peel hydrolyzate can be gallic acid and ellagic acid as well as flavogalol. Tannins are punicalagin (2,3-(S)-hexahydroxydiphenoyl-4,6-(S,S)-galagyl-D-glucose; α-punicalagin; β-punicalagin) and punicalin (4,6-(S,S)-galagyl-D-glucose) and are responsible for obtaining yellow shades.

Onion peels (*Allium cepa* L.) are rich in flavonoids, and the most common is quercetin, followed by quercetin glucoside and kaempferol. Due to their chemical structure, flavonoids are the most common plant dyes of yellow shades. The natural plant dyes used in this paper belong to the group of acid-wetting dyes. Therefore, in the acid medium in dyeing process, ionization occurs on imide, carbonyl and hydroxyl groups, and depending on the raw material composition of the textile material, the choice of mordants (metal salts) and the chemical structure of the natural dye, metal complexes of different colours are formed. The obtained dyes and their properties are the result of the formation of a ligand: fibre-metal ion-natural dye (Figure 3) [34].

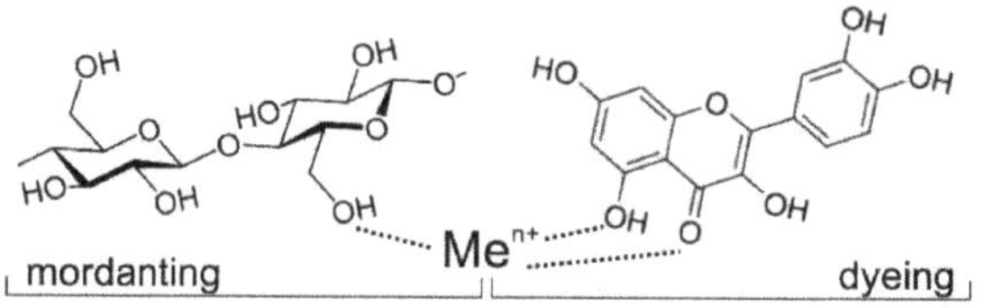

Figure 3. Schematic image of ligand formation: fibre—metal ion—natural dye.

After dyeing, the samples were spectrophotometrically measured, and the analysis of objective colour parameters was performed. The results of the coordinate placement of the obtained colours are shown in a*/b* colour space, with a graphical representation of the objective values of the basic colour parameters: lightness (L*), chroma (C*) and hue (h°), in Figure 4.

From the colour coordinates of the cotton yarns dyed with onion peel (Figure 4), it is evident that the raw cotton yarn dyed in an aqueous extract of onion peel, achieved yellow hue (h° = 73.52) of high lightness (L* = 81.60) and low chroma (C* = 15.80). With the addition of mordants, the hue changes, but the most significant change occurred was the increase in chroma value. The sample treated with aluminium salts have chroma C* = 37.43 and with copper C* = 23.30. With lightness being L* = 68.81 and 63.67, respectively, the increase in colour intensity can be confirmed, although, given that it is a yellow hue, the ratio of chroma to lightness is still insufficient to give a clear, chromatic colour. As can be seen from the sample images (Table 4), the achieved shades are in the range of brownish-yellow-orange. The sample pre-treated with ferrum salts, as it was expected, achieved the most achromatic shade with the lowest chroma C* = 23.56.

Even more emphasized achromatic shade with lower chroma, is achieved for the scoured sample treated with ferrum salt (C* = 14.14), while for the rest of the scoured samples, the similar relationship of chroma C* and lightness L* have been achieved with hue ranging also in yellow-orange spectrum.

In chemically bleached cotton yarn, all natural pigments have been removed, so the substrate is whiter, giving the hues cleaner, lighter and more yellow after the dyeing. When mordant is added, the hue of the samples changes slightly (h° = 74.77–77.90) but it stays in yellow/yellow-orange spectrum. The sample treated with aluminium has the highest chroma value (C* = 33.53) which in relation to lightness (L* = 74.41) gives the highest colour intensity, although still in chromatic-achromatic area. For the scoured samples as well the most achromatic shade is achieved for the sample treated with ferrum salts.

In mercerized samples, a positive effect of mercerization on the absorbance and affinity of cotton fibre to dyes is observed. Even for mordant-free samples, a higher chroma value is achieved compared to non-mercerized samples, while for samples treated with mordants (even for samples treated with ferum salt) the highest intensity, arising from the specific lightness/chroma relationship is shown. Furthermore, the results show that for the samples treated with slack mercerization, a higher colour intensity is achieved compared to samples mercerized with tension.

The photographs of dyed samples regarding the cotton pre-treatments and mordant usage are shown in Table 5.

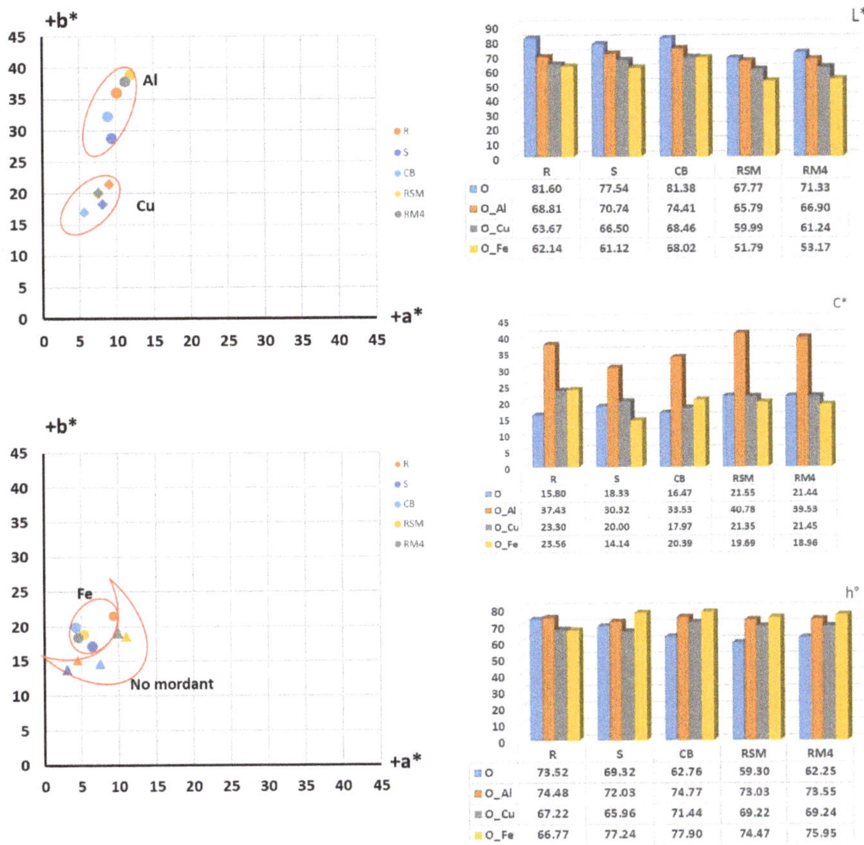

Figure 4. Colour analysis in the a*/b* colour space and objective results of lightness (L*), schroma (C*) and hue (h°) of cotton yarns dyed with onion peel.

Table 5. Cotton yarns dyed with onion peel.

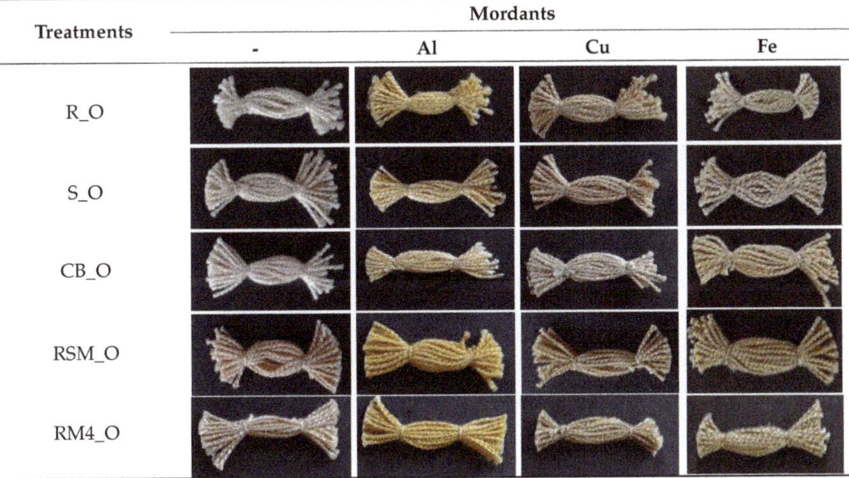

Treatments	Mordants			
	-	Al	Cu	Fe
R_O				
S_O				
CB_O				
RSM_O				
RM4_O				

For samples dyed with natural dye extracted from pomegranate peel, the equal analysis is performed and the results of positioning the obtained colourations in a*/b* colour space with numerical evaluation of the main colour parameters; lightness (L*), chroma (C*) and hue (h°) is shown in Figure 5.

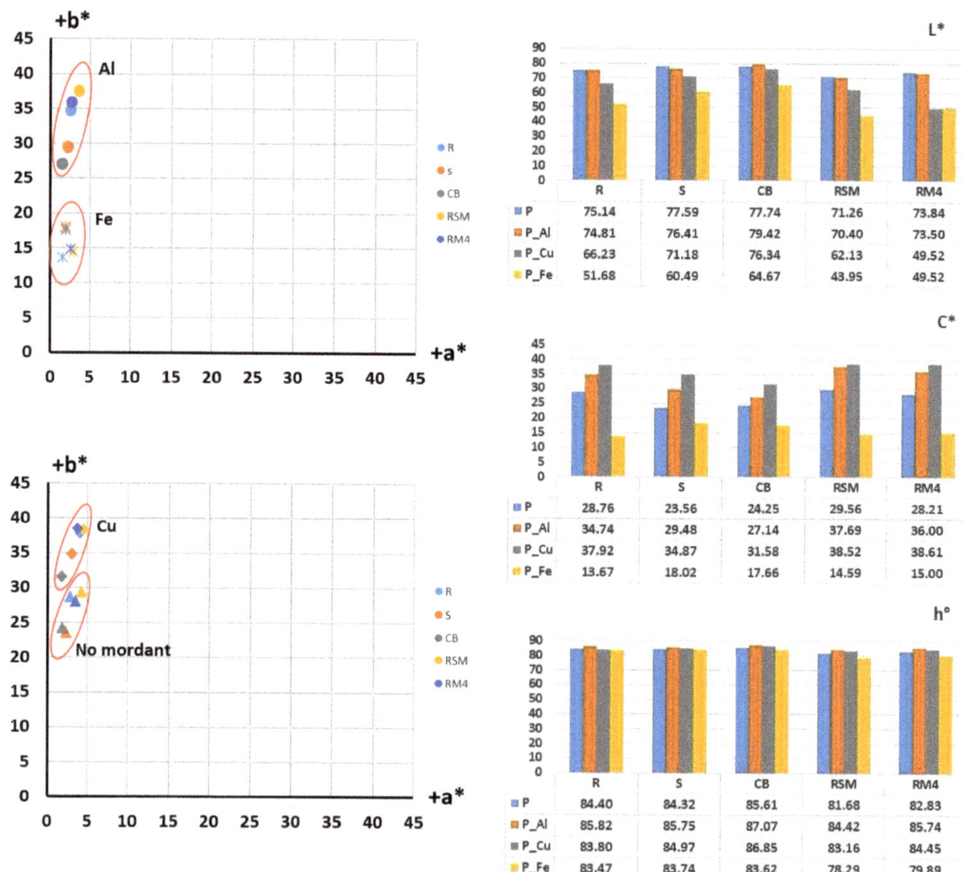

Figure 5. Colour analysis in the a*/b* colour space and objective results of lightness (L*), schroma (C*) and hue (h°) of cotton yarns dyed with pomegranate peel.

In general, a more pronounced yellow colour hue is achieved for samples dyed with pomegranate peel natural dye, compared to samples dyed with onion peel natural dye. Furthermore, higher chroma (C*) for non-mordanted samples is achieved compared to samples dyed with onion extracted dye. This arises from the fact that the pomegranate is a tannin dyestuff, and tannin has a property of behaving as a mordant. It is also important to mention that the pomegranate dye is a natural substantive dye which makes it suitable for cotton.

As for the samples treated with mordants, a similar influence on dye absorption and achieved relationship of chroma (C*) and lightness (L*) compared to the onion dyed samples can be seen. The sample treated with copper has the highest chromaticity (C* = 37.92), while the sample treated with ferrum salts has the lowest chroma value (C* = 13.67), which in relation to the lowest lightness (L* = 51.68) correspond to the visual appearance of the sample being dark greyish and the most achromatic.

The highest colour depth (which is inferred from the specific ratio of lightness and chroma) was obtained for the sample treated by slack mercerization and mordanted with copper salts.

The brightest shades are achieved for mercerized samples mordanted with aluminium salt. Although the value of chroma is not the highest for these samples, in a specific relationship with higher lightness, and given the nature of the yellow colour, visually, the appearance of a colour of greater brilliance is obtained (Table 6).

Table 6. Cotton yarns dyed with pomegranate peel.

Treatments	Mordants			
	-	Al	Cu	Fe
R_P				
S_P				
CB_P				
RSM_P				
RM4_P				

Contrary to theoretical expectations, the highest colour intensity was not obtained for bleached samples, although bleaching should improve colour clarity and brightness. Lower chroma relative to higher lightness resulted in a visually pastel, bright, unsaturated colour.

After colorimetric analysis of the dyed samples, a wash fastness test was performed. Five wash cycles were performed, and the results are shown after 1st and 5th cycles by the objective value of the total colour difference calculated according to the CIE76 system as well as by the value of grey scale grading (Tables 7 and 8).

Samples treated with onion peel after the 1st and 5th wash cycle show low colour fastness, which was to be expected due to the chemical construction of the natural dye extracted from onion peel. After the 1st wash cycle, a smaller colour difference (DE) is visible than after the 5th wash cycle. The smallest colour difference is obtained for the sample pre-treated by slack mercerization, mordanted with $CuSO_4$ ($DE_{RSM\_O\_Cu} = 1.49$). This is to be expected, because during free mercerization, the crystal lattice of cellulose changes. There is also an increase in amorphous areas in the cellulose that are responsible for higher dye binding. Grey scale values after 1st wash cycle, for most samples range between 3 and 4, which indicate rather satisfactory stability of natural dyes during washing, but after the 5th wash cycle, the wash fastness ratings decrease significantly, while DE increases. This indicates that during the 5th wash cycle, a significant amount of dye is removed from the yarns (Table 7).

Table 7. Colour fastness to washing after 1st and 5th cycles of washing of cotton yarns dyed with onion peel.

(a) Objective evaluation	(b) Grey scale evaluation		
	Labels	1st	2nd
	1	4	3
	2	3	1
	3	3	2–3
	4	3–4	2
	5	3–4	2–3
	6	3	1
	7	3–4	2
	8	3	2
	9	3	2
	10	1–2	1
	11	2	1
	12	3–4	1–2
	13	3	2
	14	3	1–2
	15	4	2
	16	2	1–2
	17	3	2–3
	18	3–4	1–2
	19	2	1
	20	3–4	1

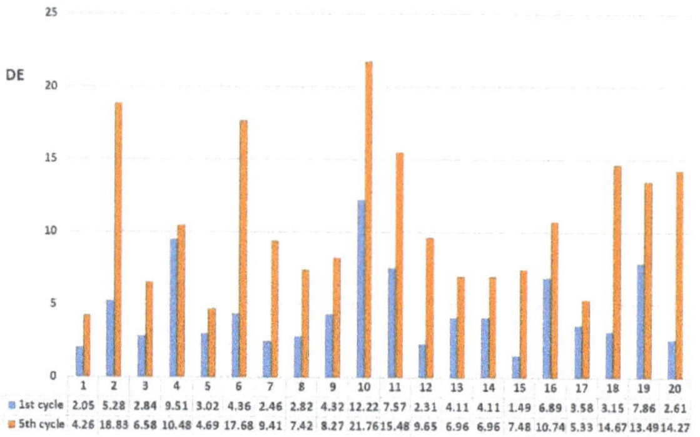

Objective evaluation (DE) — bar chart:

Label	1	2	3	4	5	6	7	8	9	10	11	12	13	14	15	16	17	18	19	20
1st cycle	2.05	5.28	2.84	9.51	3.02	4.36	2.46	2.82	4.32	12.22	7.57	2.31	4.11	4.11	1.49	6.89	3.58	3.15	7.86	2.61
5th cycle	4.26	18.83	6.58	10.48	4.69	17.68	9.41	7.42	8.27	21.76	15.48	9.65	6.96	6.96	7.48	10.74	5.33	14.67	13.49	14.27

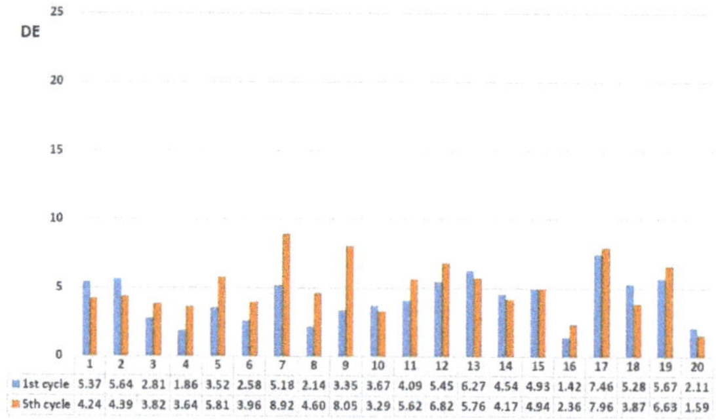

1 (R_O); 2 (R_O_Al); 3 (R_O_Cu); 4 (R_O_Fe); 5 (S_O); 6 (S_O_Al); 7 (S_O_Cu); 8 (S_O_Fe); 9 (CB_O); 10 (CB_O_Al); 11 (CB_O_Cu); 12 (CB_O_Fe); 13 (RSM_O); 14 (RSM_O_Al); 15 (RSM_O_Cu); 16 (RSM_O_Fe); 17 (RM4_O); 18 (RM4_O_Al); 19 (RM4_O_Cu); 20 (RM4_O_Fe)

Table 8. Colour fastness to washing after 1st and 5th cycles of washing of cotton yarns dyed with pomegranate peel.

(a) Objective evaluation	(b) Grey scale evaluation		
	Labels	1st	2nd
	1	3	3
	2	2–3	3
	3	3–4	3
	4	4	3
	5	3	2
	6	3	3
	7	2	2
	8	4	2–3
	9	3	2
	10	3	3
	11	3	3
	12	2	2
	13	2–3	2–3
	14	3	3
	15	3	3
	16	4	3–4
	17	2	3
	18	2–3	3
	19	2–3	2–3
	20	4–3	4

Objective evaluation (DE) — bar chart:

Label	1	2	3	4	5	6	7	8	9	10	11	12	13	14	15	16	17	18	19	20
1st cycle	5.37	5.64	2.81	1.86	3.52	2.58	5.18	2.14	3.35	3.67	4.09	5.45	6.27	4.54	4.93	1.42	7.46	5.28	5.67	2.11
5th cycle	4.24	4.39	3.82	3.64	5.81	3.96	8.92	4.60	8.05	3.29	5.62	6.82	5.76	4.17	4.94	2.36	7.96	3.87	6.63	1.59

1 (R_O); 2 (R_O_Al); 3 (R_O_Cu); 4 (R_O_Fe); 5 (S_O); 6 (S_O_Al); 7 (S_O_Cu); 8 (S_O_Fe); 9 (CB_O); 10 (CB_O_Al); 11 (CB_O_Cu); 12 (CB_O_Fe); 13 (RSM_O); 14 (RSM_O_Al); 15 (RSM_O_Cu); 16 (RSM_O_Fe); 17 (RM4_O); 18 (RM4_O_Al); 19 (RM4_O_Cu); 20 (RM4_O_Fe)

Samples dyed with pomegranate peel showed better wash fastness compared to samples dyed with onion peel. Colour differences after the 1st wash cycle are in range from DE 1.42 (RSM_P_Fe) to 7.46 (RM4_P). After the 5th wash cycle, a minimal increase in total colour difference value (DE) is obtained, which indicate that in the 1st wash cycle, the unbounded dye was removed. The results of total colour difference (DE) obtained after the 5th wash cycles show that the optimal amount of dyestuff stayed bounded to the cotton fibre showing satisfactory wash fastness property. This is confirmed by the ratings in grey scale. The grades did not decrease drastically and did not decrease at all after the 5th wash cycle, which indicates satisfactory colour durability after the 5th wash cycle (Table 8).

Furthermore, the light fastness of obtained colorations was tested. Tables 9 and 10 show the light fastness results of samples dyed with onion peel (Table 9) and pomegranate peel (Table 10) measured on Xenotest 440. The results are shown as objective numerical evaluation of total colour difference (DE) calculated according to CIE76 equation, as well as blue scale grades. The yarns dyed with pomegranate peel showed better light fastness. The colour difference (DE) after 41:10 h of light exposure of yarns dyed with pomegranate peel have ranges from 0.49 (R_P_Fe) to 7.35 (CB_P_Al), while yarns dyed with onion peel have higher colour differences ranging from 0.40 (RSM_O_Cu) to 18.81 (CB_O_Al). A significant influence of mordants on the results of light fastness was observed. Yarns dyed with onion mordanted with copper obtained better light fastness than other yarns. All yarns dyed with pomegranate peel also have the same trend. The yarns dyed with pomegranate peel mordanted with iron obtained better light fastness. An increase in colour difference (DE) was observed when using aluminium mordant for yarns dyed with both onion and pomegranate peel, which indicates a lower light fastness. The blue scale ratings confirm higher light fastness of yarns dyed with pomegranate peel dye, as well as a positive effect of mordanting.

Table 9. Light fastness of cotton yarns dyed with onion peel.

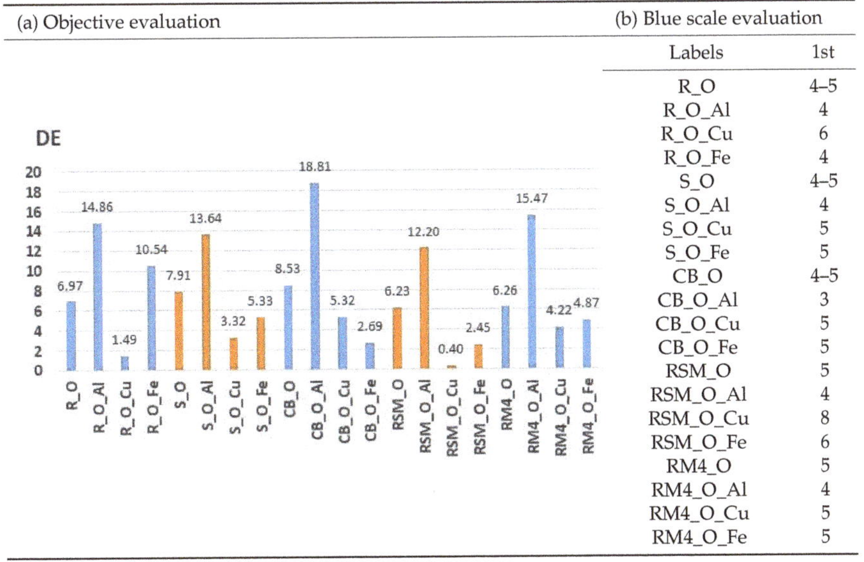

(a) Objective evaluation	(b) Blue scale evaluation	
	Labels	1st
	R_O	4–5
	R_O_Al	4
	R_O_Cu	6
	R_O_Fe	4
	S_O	4–5
	S_O_Al	4
	S_O_Cu	5
	S_O_Fe	5
	CB_O	4–5
	CB_O_Al	3
	CB_O_Cu	5
	CB_O_Fe	5
	RSM_O	5
	RSM_O_Al	4
	RSM_O_Cu	8
	RSM_O_Fe	6
	RM4_O	5
	RM4_O_Al	4
	RM4_O_Cu	5
	RM4_O_Fe	5

Table 10. Light fastness of cotton yarns dyed with onion peel.

(a) Objective evaluation	(b) Blue scale evaluation	
	Labels	1st
	R_O	5
	R_O_Al	5
	R_O_Cu	5
	R_O_Fe	8
	S_O	5
	S_O_Al	4–5
	S_O_Cu	5
	S_O_Fe	6
	CB_O	6
	CB_O_Al	4–5
	CB_O_Cu	5
	CB_O_Fe	7
	RSM_O	5
	RSM_O_Al	5
	RSM_O_Cu	6
	RSM_O_Fe	5
	RM4_O	5
	RM4_O_Al	5
	RM4_O_Cu	5
	RM4_O_Fe	5–6

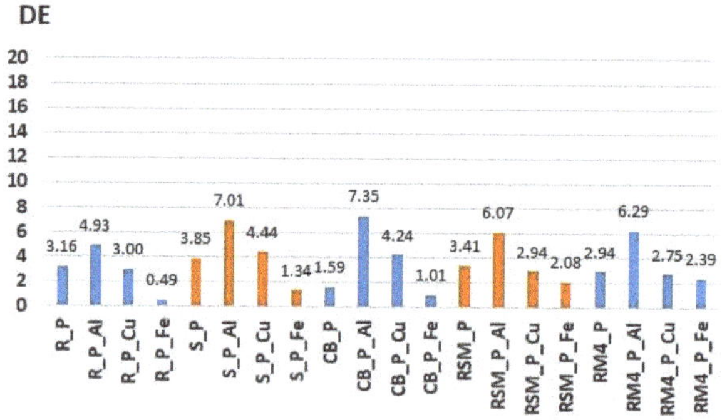

The objective evaluation chart (DE) shows the following values by label: R_P 3.16; R_P_Al 4.93; R_P_Cu 3.00; R_P_Fe 0.49; S_P 3.85; S_P_Al 7.01; S_P_Cu 4.44; S_P_Fe 1.34; CB_P 1.59; CB_P_Al 7.35; CB_P_Cu 4.24; CB_P_Fe 1.01; RSM_P 3.41; RSM_P_Al 6.07; RSM_P_Cu 2.94; RSM_P_Fe 2.08; RM4_P 2.94; RM4_P_Al 6.29; RM4_P_Cu 2.75; RM4_P_Fe 2.39.

4. Conclusions

Tests conducted in this paper confirm that, although natural dyes generally have a lower affinity for cellulose fibres, satisfactory dye quality results can be achieved. Furthermore, the positive effect of mercerization on the absorption and binding of dyes was confirmed, both in onions and pomegranates dyeing. Even without additional pre-treatment with mordants, certain colour intensity is obtained compared to untreated non-mercerized samples. However, in order to achieve more intense coloration, the necessity of applying mordants was confirmed. Optimal colour intensity results for the natural onion dye were obtained with aluminium salt as mordants, while for the pomegranate dye, the copper salts were confirmed as optimal.

Contrary to expectations, chemical bleaching did not improve the achievement of more intense coloration, but very light, pastel shades of low intensity were obtained.

Yarns treated with onion peel, after the 1st and 5th wash cycles show low colour fastness, which was to be expected due to the chemical constitution of the natural dye extracted from onion peel. For yarns dyed with pomegranate peel after the 5th wash cycle, a minimal increase in the total colour difference value (DE), compared to the 1st wash cycle was obtained, indicating a better wash fastness for yarns dyed with pomegranate peel.

As for the light fastness, for yarns dyed with pomegranate peel, better light fastness was obtained. It must be emphasized that the selection of mordant is also an important factor in light fastness. Thus, for the yarns dyed with onion peel mordanted with copper and yarns dyed with pomegranate peel mordanted with ferrum, satisfactory light fastness was obtained. For yarns mordanted with aluminium, the lowest light fastness on all dyed yarns were obtained.

Author Contributions: Conceptualization, L.B.; methodology, L.B. and M.I.G.; experimental part, A.M.; measurements, A.M.; supervision, L.B.; writing—original draft preparation, L.B. and M.I.G.; writing—review and editing, M.I.G. All authors have read and agreed to the published version of the manuscript.

Funding: The work was funded by the research support program from the Ministry of Science, Education and Sports of Republic Croatia and the University of Zagreb. Support was granted to the

Faculty of Textile Technology for the research topic Multifunctional properties of textile materials treated with natural dyes, for year 2022.

Institutional Review Board Statement: Not applicable.

Informed Consent Statement: Not applicable.

Data Availability Statement: The data presented in this study are available on request from the corresponding author.

Conflicts of Interest: The authors declare no conflict of interest.

Sample Availability: Samples of the compounds are available from the authors. A paper recommended by the 14th Scientific and Professional Symposium Textile Science and Economy, The University of Zagreb Faculty of Textile Technology.

References

1. Samanta, A.K.; Agarwal, P. Application of natural dyes on textiles. *Indian J. Fibre Text. Res.* **2009**, *4*, 384–399.
2. Glogar, M.; Tancik, J.; Brlek, I.; Sutlovic, A.; Tkalec, M. Optimisation of process parameters of Alpaca wool printing with Juglans regia natural dye. *Coloration Technol.* **2020**, *136*, 188–201.
3. Parac-Osterman, Đ. Bojadisanje vune prirodnim bojilima u svjetlu etnografske baštine Like. *Tekstil* **2001**, *50*, 339–344.
4. Kumar, R.; Ramratan; Kumar, A. To study natural herbal dyes on cotton fabric to improving the colour fastness and absorbency performance. *J. Text. Eng. Fash. Technol.* **2021**, *7*, 51–56.
5. Sutlović, A.; Brlek, I.; Ljubić, V.; Glogar, M.I. Optimization of Dyeing Process of Cotton Fabric with Cochineal Dye. *Fibers Polym.* **2020**, *21*, 555–563. [CrossRef]
6. Samanta, A.K.; Singhee, D.S. Application of Single and Mixture of Selected Natural Dyes on Cotton Fabric: A Scientific Approach. *Colourage* **2003**, *10*, 29–42.
7. Ferreira, E.S.B. New Approaches towards the Identification of Yellow Dyes in Ancient Textiles. Ph.D. Thesis, The University of Edinburgh, Edinburgh, UK, 2002.
8. Ferreira, E.S.B.; Hulme, A.N.; McNab, H.; Quye, A. The natural constituent of historical textile dyes. *Chem. Soc. Rev.* **2004**, *33*, 329–336. [CrossRef]
9. Ferreira, E.S.B.; Quye, A.; McNab, H.; Hulme, A.N. Photo-oxidation Products of Quercetin and Morin as Markers for the Characterisation of natural Yellow Dyes in Ancient Textiles. *Dye. Hist. Archaeol.* **2002**, *18*, 63–72.
10. Iqbal, Z.; Liaqat, L. Dyeing Properties of Natural Dye Extracted from Punica Granatum Bark. *World J. Pharm. Res.* **2018**, *18*, 38–44.
11. Gupta, D.; Khare, S.K.; Laha, A. Antimicrobial Properties of Natural Dyes Against Gram-negative Bacteria. *Coloration Technol.* **2004**, *4*, 167–171. [CrossRef]
12. Hwang, E.K.; Lee, Y.H.; Kim, H.D. Dyeing, Fastness and Deodorizing Properties of Cotton, Silk and Wool Fabrics Dyed with Gardenia, Coffee Sludge, Cassia Tora. L. and Pomegranate Extracts. *Fibers Polym.* **2008**, *3*, 334–340. [CrossRef]
13. Davulcu, A.; Benli, H.; Şen, Y.; Bahtiyari, M.I. Dyeing of Cotton with Thyme and Pomegranate Peel. *Cellulose* **2014**, *21*, 4671–4680. [CrossRef]
14. Kulkarni, S.S.; Gokhale, A.V.; Bodake, U.M.; Pathade, G.R. Cotton Dyeing with Natural Dye Extracted from Pomegranate (Punica granatum) Peel. *Univers. J. Environ. Res. Technol.* **2011**, *1*, 135–139.
15. Satyanarayana, D.N.V.; Ramesh Chandra, K. Dyeing Of Cotton Cloth with Natural Dye Extracted From Pomegranate Peel and its Fastness. *Int. J. Eng. Sci. Res. Technol.* **2013**, *2*, 2664–2669.
16. Adeel, S.; ASli, S.; Bhatti, A.I.; Zsila, F. Dyeing of Cotton Fabric using Pomegranate (Punica granatum) Aqueous Extract. *Asian J. Chem.* **2009**, *21*, 3493–3499.
17. Silva, M.G.; Barros, M.A.S.D.; Almeida, R.T.R.; Pilau, E.J.; Pinto, E.; Ferreira, A.J.S.; Vila, N.T.; Soares, G.; Santos, J.G. Multifunctionalization of Cotton with Onion Skin Extract. In Proceedings of the IOP Conference Series Materials Science and Engineering, Istanbul, Turkey, 20–22 June 2018; Volume 1, pp. 1–6.
18. Sadeghi-Kiakhani, M.; Tehrani-Bagha, A.R.; Gharanjig, K.; Hashemi, E. Use of pomegranate peels and walnut green husks as the green antimicrobial agents to reduce the consumption of inorganic nanoparticles on wool yarns. *J. Clean. Prod.* **2019**, *231*, 1463–1473. [CrossRef]
19. Karaboyaci, M. Recycling of rose wastes for use in natural plant dye and industrial applications. *J. Text. Inst.* **2014**, *105*, 1160–1166. [CrossRef]
20. Zuin, V.G.; Ramin, L.Z. Green and Sustainable Separation of Natural Products from Agro-Industrial Waste: Challenges, Potentialities, and Perspectives on Emerging Approaches. *Top. Curr. Chem.* **2018**, *376*, 3. [CrossRef]
21. Karuppuchamy, A.; Annapoorani, G. Natural dye extraction from agro-waste and its application on textiles. *Asian Dye.* **2019**, *16*, 35–39.
22. Choi, I.S.; Cho, E.J.; Moon, J.-H.; Bae, H.-J. Onion skin waste as a valorization resource for the by-products quercetin and biosugar. *Food Chem.* **2015**, *188*, 537–542. [CrossRef]

23. Singhee, D. Review on Natural Dyes for Textiles from Wastes. In *Chemistry and Technology of Natural and Synthetic Dyes and Pigments*, 3rd ed.; Samanta, A.K., Awwad, N.S., Algarni, H.M., Eds.; IntechOpen Limited: London, UK, 2020; Chapter 4.
24. Reddy, J.P.; Rhim, J.W. Extraction and Characterization of Cellulose Microfibers from Agricultural Wastes of Onion and Garlic. *J. Nat. Fibers* **2018**, *15*, 465–473. [CrossRef]
25. Ferdush, J.; Nahar, K.; Akter, T.; Ferdoush, M.J.; Iqbal, S.F. Effect of Hydrogen Peroxide Concentration on 100% Cotton Knit Fabric Bleaching. *Eur. Sci. J.* **2019**, *15*, 254–263. [CrossRef]
26. Abdel-Halim, E.S. Simple and economic bleaching process for cotton fabric. *Carbohydr. Polym.* **2012**, *88*, 1233–1238. [CrossRef]
27. Tarbuk, A.; Grancarić, A.M.; Leskovac, M. Novel cotton cellulose by cationization during the mercerization process—Part 1: Chemical and morphological changes. *Cellulose* **2014**, *21*, 2167–2179. [CrossRef]
28. Pušić, T.; Grancarić, A.M.; Soljačić, I. The influence of bleaching and mercrisation of cotton on then change of electrokinetic potential. *Vlak. A Text.* **2001**, *8*, 121–124.
29. Bristi, U.; Pias, A.K.; Lavlu, F.H. A Sustainable process by bio- scouring for cotton knitted fabric suitable for next generation. *J. Text. Eng. Fash. Technol.* **2019**, *5*, 41–48. [CrossRef]
30. Mojsov, K. Enzyme scouring of cotton fabrics: A review. *Int. J. Mark. Technol.* **2012**, *2*, 256–275.
31. Dinand, E.; Vignon, M.; Chanzy, H.; Heux, L. Mercerization of Primary Wall Cellulose and its Implication for the Conversion of Cellulose I to cellulose II. *Cellulose* **2002**, *9*, 7–18. [CrossRef]
32. Kim, S.I.; Lee, E.S.; Yoon, H.S. Mercerization in degassed sodium hydroxide solution. *Fiber Polym.* **2006**, *7*, 186–190. [CrossRef]
33. Sutlović, A.; Parac-Osterman, Đ.; Đurašević, V. Croatian Traditional Herbal Dyes For Textile Dyeing, International Interdisciplinary. *J. Young Sci. Fac. Text. Technol.* **2011**, *1*, 65–69.
34. Sutlovic, A.; Glogar, M.I.; Beslic, S.; Brlek, I. Natural textile dyes—A contribution to creativity and sustainability (Prirodna bojila za tekstil—Doprinos kreativnosti i odrzivosti). *Tekstil* **2020**, *69*, 1–10.
35. Hofenk de Graaff, J. *The Colourful Past: Origins, Chemistry and Identification of Natural Dyestuffs*; Archetype Publications Ltd.: London, UK, 2004.
36. Peran, J.; Ercegović Ražić, S.; Sutlović, A.; Ivanković, T.; Glogar, M. Oxygen Plasma Pre-Treatment Improves Dyeing and Antimicrobial Properties of Wool Fabric Dyed with Natural Extract from Pomegranate Peel. *Coloration Technol.* **2020**, *136*, 177–187.

Article

Natural Dyeing of Modified Cotton Fabric with Cochineal Dye

Ivana Čorak, Iva Brlek *, Ana Sutlović and Anita Tarbuk

Department of Textile Chemistry and Ecology, University of Zagreb Faculty of Textile Technology, HR-10000 Zagreb, Croatia; ivana.corak@ttf.unizg.hr (I.Č.); ana.sutlovic@ttf.unizg.hr (A.S.); anita.tarbuk@ttf.unizg.hr (A.T.)
* Correspondence: iva.brlek@ttf.unizg.hr

Abstract: Natural dyes are not harmful to the environment owing to their biodegradability. For dye application to textiles, salts are necessary as mordant or electrolytes and make an environmental impact. In this paper, the influence of cationization during mercerization to the dyeing of cotton fabric with natural dye from *Dactylopius coccus* was researched. For this purpose, bleached cotton fabric as well as fabric cationized with Rewin OS was pre-mordanted using iron(II) sulfate heptahydrate ($FeSO_4 \cdot 7H_2O$) and dyed with natural cochineal dye with and without electrolyte addition. For the characterization of surface changes after cationization, an electrokinetic analysis on SurPASS was performed and compared to pre-mordanting. For determination of dye exhaustion, the analysis of dye solution was performed on a UV/VIS spectrophotometer Cary 50 Solascreen. Spectrophotometric analysis was performed using a Datacolor 850 spectrophotometer, measuring remission "until tolerance" and the whiteness degree, color parameters, color depth (K/S), and colorfastness of dyed fabric were calculated. Levelness was determined by visual assessment. Cationized cotton fabrics showed better absorption and colorfastness. Pre-mordanting and cationization showed synergism. The electrolytes improved the process of dye absorption. However, when natural dyeing was performed on cotton fabric cationized during mercerization, similar chromacity, uniform color, and colorfastness were achieved with and without electrolyte, resulting in pure purple hue of cochineal. For achieving a violet hue, pre-mordanting with Fe-salt was needed. Therefore, salt can be reduced or even unnecessary, which makes this process of natural dyeing more environmentally friendly.

Keywords: cotton fabric; cationization during mercerization; pre-mordanting; dyeing; natural cochineal dye

Citation: Čorak, I.; Brlek, I.; Sutlović, A.; Tarbuk, A. Natural Dyeing of Modified Cotton Fabric with Cochineal Dye. *Molecules* **2022**, *27*, 1100. https://doi.org/10.3390/molecules27031100

Academic Editor: Baljinder Kandola

Received: 28 December 2021
Accepted: 5 February 2022
Published: 7 February 2022

Publisher's Note: MDPI stays neutral with regard to jurisdictional claims in published maps and institutional affiliations.

1. Introduction

The application of natural dyes is currently under investigation owing to the multifunctional properties of natural dyes, i.e., inhibition of the growth of some pathogenic bacteria, good protection against UV radiation, and others [1–5]. Natural dyes are easily renewable or derived from waste raw materials of plant origin and are therefore also environmentally friendly [2–10]. It is well known that environmental parameters, i.e., chemical (COD) and biochemical (BOD) oxygen demands, are 65% lower than if synthetic dye were used [6,7]. This was confirmed by the authors after natural dyeing with ash bark extract [8].

Dactylopius coccus is an insect from which natural dye can be extracted. This insect most often lives in tropical and subtropical areas of Mexico and Central America and the northern Andes in South America. It takes 155,000 insects to produce one kilogram of cochineal dye [11–14]. Extracted dye is named cochineal or carmine due to its chemical structure, which is carminic acid [2,15–17]. Cochineal dye—C.I. Natural Red 4, 75470, by Naturex—is chosen for this research because of good regulation status. This dye is a color additive permitted in food and compliant with the purity criteria set by European regulation [18]. Except from meeting strict toxicological parameters for the food industry, one of the main advantages is its ecological acceptability, i.e., good biodegradability. According to the FTIR-ATR spectrum of this specific dye, absorption bands of the main components of carminic acid correspond to anthraquinone compounds. Typical absorption peaks indicate the presence of hydroxyl (–OH), carbonyl (C=O), and carboxyl (–COOH) groups [19–21].

During the dyeing process, metal salts are used because most of the natural dyestuffs are unable to form strong bonds with fibers. This process is called mordanting and it improves the fastness properties of dyed fabrics. Unfortunately, it can change color properties and the resulting hue, and can have negative environmental impact [7,19,22–24]. However, it has been shown that these disadvantages can be avoided. Difference in hue can expand the color palette for creation of specific design, whilst environmental impact can be lowered by optimization of mordant concentration considering the dye and textile material. In last two decades, different concentrations of metal salts were researched from low concentrations of 0.5% owf (over the weight of the fabric) to abnormally high concentrations of 10% [25], 20% [26,27], and even 10–100% owf (5–50 g/L, 1:20 OK) [28,29]. For cochineal dye, Michael et al. suggested 1.6–5% as optimal for cotton depending on the metal salt (KAl(SO$_4$)$_2$, FeSO$_4$, and CuSO$_4$) [30], and Brukner et al. optimized KAl(SO$_4$)$_2$ and CuSO$_4$ for wool and synthetic fibers [31]. Sutlović et al. [19,32] researched different mordants for cochineal dye (KAl(SO$_4$)$_2$·12H$_2$O; CuSO$_4$ 5H$_2$O; FeSO$_4$·7H$_2$O) in a concentration gradient of 0, 0.5, 3, 5, and 10% owf. It was confirmed that the use of mordant affects the color depth obtained, and that metal concentration of 3–5% owf is satisfactory [19]. The use of higher concentrations of metals can have negative impact to the environment and affect human health [7,19]. From the aspect of the color, more chromatic color shades with color hue in the range 333.77–339.03° were obtained with KAl(SO$_4$)$_2$, whereas the near achromatic shades with hue range 318.41–332.81° were obtained with FeSO$_4$. This is indicating the role of mordant agents in achieving a wider color palette of different shades [19]. Additionally, in previous research [8], the amount of iron and copper after mordanting woolen substrate with FeSO$_4$·7H$_2$O and CuSO$_4$·5H$_2$O was studied. The residual amount after mordanting with iron (Fe^{2+}) was 6.20 mg/L for 0.5% owf and 6.87 mg/L for 2%, which is within the EU tolerance limit Fe < 10 mg/L. Because of its violet hue and these environmental parameters, FeSO$_4$ was used as the mordant in this research.

There are more recent papers concerning the improvement of cotton natural dyeing by plasma, ultrasound, gamma ray irradiation, chitosan, cationic agents, which can help to reduce the addition of mordant and/or electrolyte in the dyeing bath [33–40]. Haji et al. studied plasma treatment and subsequent attachment of chitosan biopolymer for surface functionalization of wool and cotton [37,38], and Peran et al. [36] on wool. Plasma treatment functionalized surface and cationic chitosan contributed to improved dyeability with natural dyes. Application of cationic agents, including full cationization of cotton, leads to salt-free dyeing [36–38]. Haddar et al. [33,34] used three commercial cationic agents, Croscolor DRT, Croscolor CF, and Stabifix NCC, for cationization as pretreatment to the dyeing process. Stabifix NCC and Croscolor DRT enhanced exhaustion and fixation of hibiscus extract and fennel leaf extract, respectively. Additionally, the cationization process significantly lowered COD and BOD values [34]. Cationization, a modification with amines and quaternary ammonium compounds, results in a change of fiber surface charge, thus reducing or even eliminating the usage of electrolytes in the dyeing process, and is called "salt free" [33,34,40–53]. The cationization agents and techniques have been intensively researched during the last few decades because of this environmental benefit [48–52]. Most common short-chained compounds are epihalohydrins, 2,3-epoxypropyltrimethyl ammonium chloride (EPTAC), and 3-chloro-2-hydroxypropyltrimethyl ammonium chloride (CHPTAC), which give results within 24 h, but time can be reduced to 5 h [47]. Commercial cationic compounds are usually long-chain compounds with polyammonium bonds. Cationic agents are usually applied by exhaustion and padding as pre- or after-treatment technique, while the cationization during mercerization was recently developed and optimized [48–53].

Beginning in 2003 [53] and developed by 2009 [51], this modification resulted in new cotton cellulose [41,42]. The process was optimized and it was proven that 5 h is sufficient for epihalohydrines, and only 1 h for long-chain commercial compounds [47,52]. The main difference between techniques of pre- or after-treatment and cationization during mercerization is the levelness after the dyeing process. When the cationization is performed in the after-treatment, it remains on the surface, helping exhaustion and fixation of the dyestuff and anionic auxiliaries, but this treatment blocks surface groups and the color is not leveled. However, if the cationization is performed during the mercerization process, new cellulose is formed, resulting in permanent modification with all benefits of mercerization and cationization [41,42,47,52]. Benefits of mercerization are well known—better strength, gloss, and sorption properties, i.e., dye sorption, as a result of conversion of the crystal lattice from cellulose I to cellulose II accompanied by the change in amorphous region of fiber due to different recrystallization [41,42,52,54–56]. Cationization results in the change of the surface charge that ensures further quality improvement, i.e., dye adsorption of 3% direct dye up to 99% without electrolyte [41,42,51] Cationization during mercerization was researched with application of different cationic agents and different characterization methods were performed (FT–IR, SEM, TGA, EKA, surface charge, etc.) [41,42,51,52]. It has been shown that the mercerization is the dominating process, so the changes that can be seen by FT–IR and SEM are contributed to mercerization, and just retained in cationization. If short-chain epihalohydrins were used, the change in cellulose was proven by EKA and TGA. The compounds that were investigated, either short or long-chain, did not have a difference in FT–IR spectrum [51,52]. Only the commercial compound, Rewin OS, if applied during mercerization, showed a weak peak at 1641 cm^{-1}, which can be attributed to amine, but it is not representative to perform an analysis [52]. However, electrokinetic analysis (EKA) can be performed since the differences in zeta potential curves are significant and the changes can be well observed. Additionally, it should be noted that cotton fabric cationized during mercerization with cationic reactive polyammonium compound Rewin OS was proven to have positive surface charge and zeta potential, suggesting similar binding to cellulose as epihalohydrins [52]. Since electrokinetics is important for the adsorption of dye anion, in this paper the EKA technique was chosen. Considering the environmental impact of this technique, it is the same as mercerization, which has been a common process in the textile industry for 130 years and has not changed significantly since then. If the cationizating agent was applied during the process, the cationic agent is bonded within the cellulose chains and etherification is complete. Therefore, the modification is permanent and uniform (leveled). In previous research AOX, TOC, and TN value were determined in fabric water extract after the cationization with different agents. Water extract of cotton fabrics before and after modifications indicate that the level of chloride, organic carbon, and nitrogen released by fabrics into the environment is below the EU limits (AOX < 500 µg Cl/L; TOC < 40 mg/L; TN < 2 mg/L) [51].

Compared to synthetic dyes, the color range of natural dyestuffs is rather limited and depends on pretreatment with mordants. For environmental reason, the dyeing of modified cotton cellulose by cationization during mercerization with natural cochineal dye was performed and compared to the usual one with mordant and electrolyte. To reach violet hue, cationized cotton was pre-mordanted as well.

2. Materials and Methods

In this research, 100% cotton fabric supplied by Čateks d.o.o. (Čakovec, Croatia) was used. Fabric was plain-woven of mass per unit area 160.8 g/m^2, yarn density of warp 35.8 threads/cm, and weft 20 threads/cm, scoured and bleached under industrial conditions.

Cotton fabric was cationized during the mercerization process with cationic reactive polyammonium compound Rewin OS (CHT-Bezema, Montlingen, Switzerland) on a jigger in a two-step procedure at room temperature. Firstly, the mercerization was performed in a bath with 24% NaOH and with 8 g/L Subitol MLF (CHT-Bezema, Switzerland) for 5 passages. Secondly, before fixation in hot water, alkaline cotton fabric was cationized in a bath containing 50 g/L Rewin OS dissolution in water (5 passages), then sealed and left for 1 h at room temperature. Afterward, fixation in hot water, neutralization in 5% acetic acid, and rinsing to neutral was performed. The fabric was air-dried.

Pre-mordanting of bleached and cationized cotton fabrics was performed in Polycolor, Mathis, LR 1:30, 50 °C, 30 min using iron(II) sulfate heptahydrate ($FeSO_4 \cdot 7H_2O$) as mordant with a concentration of 5% owf. Fabrics were rinsed with cold water after mordanting.

The dyeing process of cotton fabrics was performed using cochineal dye, C.I. Natural Red 4, 75470, by Naturex, Figure 1. It was performed in Polycolor, Mathis, with a concentration of 6% owf by the exhaustion method with the following parameters: LR 1:30, 60 min, 95 °C, at pH of distilled water (pH 6.5), with and without addition of 40 g/L NaCl as an electrolyte. After the dyeing process, fabrics were washed in cold water, soaped, and again washed. Soaping was performed with Cotoblanc SEL (CHT-Bezema, Switzerland). All samples were air-dried. The dyeing process was performed in two series.

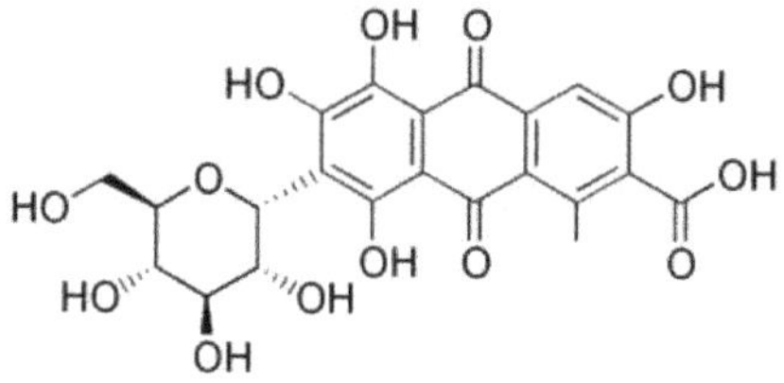

Figure 1. C.I. Natural Red 4, 75470 (Carminic acid).

Labels and treatment descriptions used to define the samples are listed in Table 1.

Table 1. Labels and description of treatments.

Label	Description of Cotton Fabric Treatment
B_	Bleached cotton fabric
_OS_	Cationized cotton with Rewin OS
_Fe_	Pre-mordanting using iron(II) sulfate heptahydrate ($FeSO_4 \cdot 7H_2O$)
C_	Cotton dyed with cochineal
… _EL	The electrolyte added in the dyeing bath
… _W	One washing cycle after dyeing

For the characterization of surface changes after cationization and pre-mordanting of cotton fabrics, electrokinetic analysis was performed. The streaming potential was measured with a SurPASS electrokinetic analyzer (Anton Paar GmbH, Graz, Austria) and the electrokinetic potential (zeta, ζ, ZP) was calculated according to the Helmholtz–Smoluchowsky equation [57]. Zeta potential was measured using an adjustable gap cell and was measured as a function of pH of the 1 mmol/L KCl, and the isoelectric point (IEP) was determined.

For the monitoring of dye exhaustion, the analysis of dye solution was performed on a Cary 50 Solascreen UV/VIS spectrophotometer (Varian, Australia). Dye exhaustion (D_{ex}) was calculated according to the following equation:

$$D_{ex} = ((D_0 - D_B)/D_0) \cdot 100 \tag{1}$$

where D_{ex} (%) is exhausted dye, D_0 (g/L) initial dye concentration, and D_B (g/L) the dye concentration in the bath at the end of the process.

For determination of color fastness, EMPA ECE reference detergent 77 without optical brightener by Testfabrics, Inc. was used. Washing was performed in Polycolor, Mathis, at 40 °C, 40 min with the program Washtest 40 using 2 g/L of detergent.

Spectral characteristics was measured using a Datacolor 850 spectrophotometer according to ISO 105-J01:1997 Textiles—Tests for colour fastness—Part J01: General principles for measurement of surface colour under illuminant D65, 8° standard observer with the specular component excluded and the UV component included. Measurement was performed at random locations on the samples from both series, using the Datacolor Tools computer program and "Measuring until tolerance" command. This means that at least 10 measurements must be made, and the results are accepted when the total color difference between each measurement is less than 0.1 ($\Delta E^* < 0.1$). The whiteness degree according to CIE (W_{CIE}) was calculated automatically according to ISO 105-J02:1997 Textiles—Tests for colour fastness—Part J02: Instrumental assessment of relative whiteness for the cationized and pre-mordanted fabrics. From the results of spectrophotometric characteristics of dyed fabrics, the color depth (K/S) and CIEL*a*b* parameters were calculated and compared with the ones after one washing cycle. The total color difference (ΔE_{CMC}) was calculated using ISO 105-J03:2009 Textiles—Tests for colour fastness—Part J03: Calculation of colour differences. The numerical value of ΔE_{CMC} within tolerance limits ($\Delta E_{CMC} \leq 2$) was used for the evaluation [58].

3. Results and Discussion

In this article, the influence of cationization during mercerization to the dyeing of cotton fabric with natural dye from *Dactylopius coccus* was researched. Cotton fabric was cationized during the mercerization process with Rewin OS, and compared to pre-mordanted with FeSO$_4$.

For the characterization of surface changes after cationization and pre-mordanting of cotton fabrics, electrokinetic analysis was performed and fabric whiteness was determined. Results are presented in Figure 2 and Table 2.

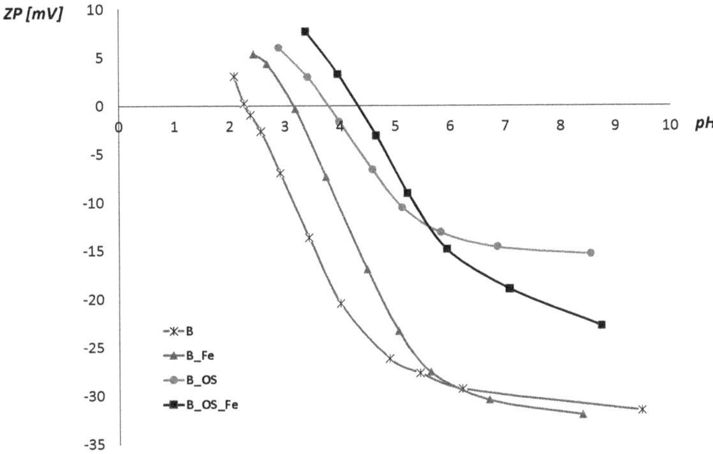

Figure 2. Electrokinetic potential of bleached (B), cationized (_OS), and pre-mordanted (_Fe) cotton fabrics vs. pH of 1 mmol/L KCl.

Table 2. Zeta potential at pH 3.5, 6.5, and 8.5, isoelectric point (IEP), and whiteness degree (W_{CIE}) of bleached, cationized, and pre-mordanted cotton fabrics.

Sample	ZP at pH 3.5/mV	ZP at pH 6.5/mV	ZP at pH 8.5/mV	IEP	W_{CIE}
B	−15.0	−29.1	−31.3	2.4	84.01
B_Fe	−5.0	−30.0	−32.1	3.2	−31.57
B_OS	2.2	−14.4	−15.0	3.8	77.81
B_OS_Fe	7.3	−17.5	−22.5	4.4	−3.49

The electrokinetic potential (zeta, ZP) vs. pH of 1 mmol/L KCl was determined on chemically bleached cotton fabrics after cationization with Rewin OS and pre-mordanting with Fe-salt. From the results shown in Figure 2 and Table 2, it can be seen that bleached cotton fabric is negatively charged in the whole pH range due to the presence of carboxyl

(–COOH) and hydroxyl (–OH) groups, having IEP at pH 2.4. Carboxyl (–COOH) and hydroxyl (–OH) groups are revealed after scouring and bleaching processes [42,56].

Pre-mordanting with iron(II) sulfate heptahydrate slightly increases negative charge in alkaline and neutral media, but in an acidic medium, positive charge of Fe-ions result in lower ZP (-5 mV). Therefore, IEP moves to 3.2 in regard to bleached cotton fabric.

The notable change in surface charge occurred in the cationization during the mercerization process. In cationized fabrics, -NH$_2$ groups are also presented besides –OH and –COOH groups, resulting in higher zeta potential (ZP = -15 mV) for alkaline and neutral media. In an acidic medium, a positive surface charge is achieved, which confirms the binding of cationic compound Rewin OS to the surface sites, moving IEP to pH 3.8. In the case of pre-mordanted cationized fabric, the same phenomenon can be observed as on bleached cotton fabric, i.e., it lowers zeta potential in neutral and alkaline medium, but increase in an acidic one. However, the values of ZP are significantly higher compared to bleached or just pre-mordanted cotton fabrics. Owing to ZP increment, IEP moves to 4.4.

As it can be seen from the results of whiteness degree presented in Table 2, all pre-treatments have influenced fabric whiteness. Chemically bleached cotton fabric has W$_{CIE}$ = 84.01. Cationization results in a slightly lower whiteness degree, W$_{CIE}$ = 77.81. By adding pre-mordanting agent iron(II) sulfate heptahydrate, the degree of whiteness evidently decreased to W$_{CIE}$ = -31.57 on noncationized cotton fabric and -3.49 on the cationized one. Negative whiteness degree indicates that fabrics changed the color, i.e., it yellowed.

Nevertheless, the results of the electrokinetic analysis indicated that the positive charge of all pre-treated cotton fabrics should result in better dyeing properties. Therefore, the dye exhaustion and color depth were determined.

For the purpose of determination of dye exhaustion, an analysis of dye solution was performed on the UV/VIS spectrophotometer. The absorption spectrum was measured (Figure 3a) and the highest absorption of cochineal dye was determined at 515 nm. Afterward, the solutions for the calibration were measured at 515 nm, and the calibration curve and equation were determined (Figure 3b). This method is based on Lambert–Beer's law, which provides the function ratio between absorbance (A) and concentration (c).

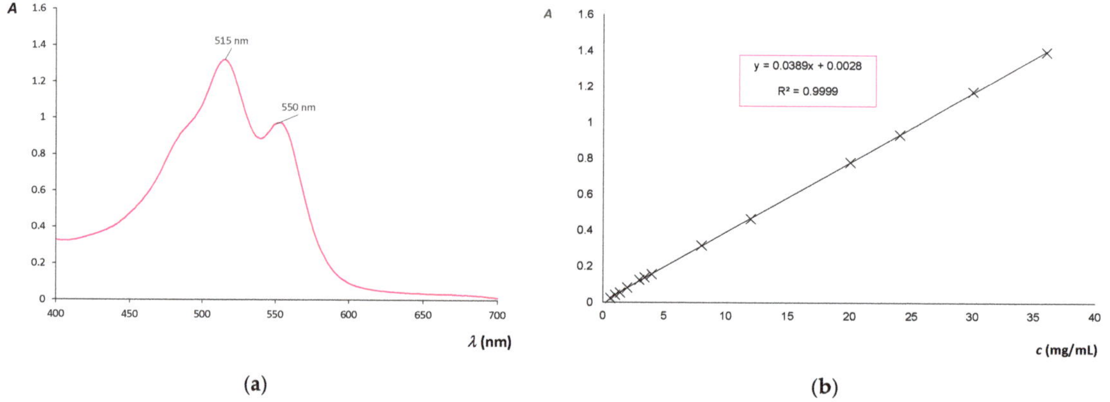

(a)   (b)

Figure 3. (**a**) Cochineal absorption spectrum. (**b**) Calibration curve for the dye concentration in regard to absorbance.

The calibration curve was used for the calculation of the exhaustion of cochineal dye after the dyeing process. The absorbance of dye solution before and after the dyeing process was measured with the UV/VIS spectrophotometer. The dye exhaustion was calculated taking into account the initial dye concentration and the dye concentration in the bath at the end of the dyeing process. The results are expressed in percent and shown in Figure 4. Bleached, noncationized cotton has exhaustion of natural cochineal dye of only 5%. With addition of electrolyte the exhaustion increases to 10.51%. Premordanting using iron(II) sulfate heptahydrate increases the exhaustion to 12.72%, and 15.02% if electrolyte was added.

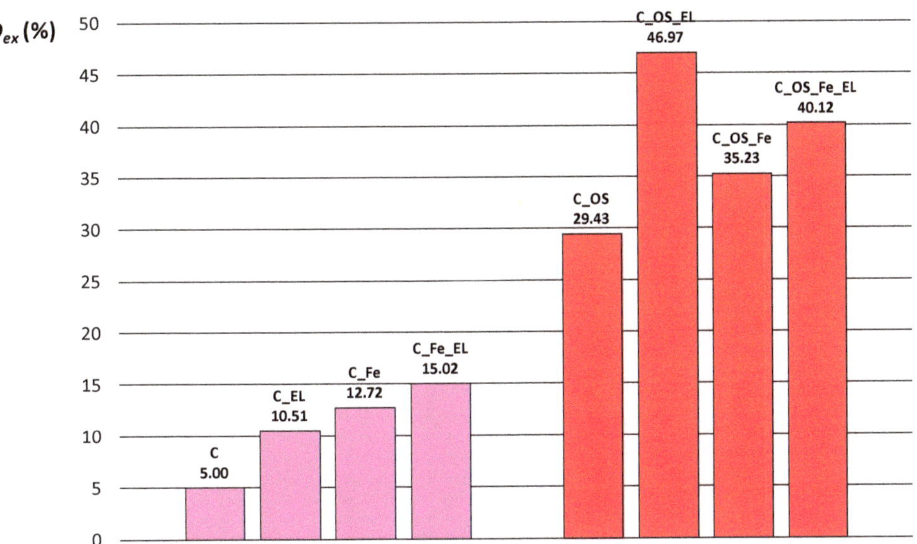

Figure 4. Dye exhaustion of differently pre-treated cotton fabrics.

It can be seen from Figure 4 that cationized cotton has significantly higher exhaustion compared to noncationized cotton fabrics, regardless of pre-mordanting or electrolyte addition. The dye exhaustion on cationized cotton is 29.43%. The electrolyte addition increase exhaustion to 46.97%. Pre-mordanting of cationized cotton fabric leads to higher exhaustion, 35.23% in regard to 29.43%. However, the exhaustion with electrolyte on pre-mordanted cationized fabric (C_OS_Fe_EL) is not higher than if not pre-mordanted. The highest exhaustion is on cationized fabric with the addition of electrolyte. The main reason for such high dye exhaustion is change of surface charge.

Spectral analysis was performed from the results of spectrophotometric values after the dyeing process and after one washing cycle for the reason of color fastness. The results of color parameters are shown in Tables 3 and 4. The color depth coefficient K/S was calculated from remission and presented as K/S maximum at 520 nm in Figure 5. Visual representation of all dyed fabrics is given in Figure 6.

Table 3. Color parameters of cochineal dyed cotton fabrics.

Sample	L*	a*	b*	C*	h°
C	94.20	0.80	2.04	2.19	68.48
C_EL	92.02	2.85	0.73	2.94	14.42
C_Fe	79.02	6.90	13.78	15.41	63.39
C_Fe_EL	78.80	6.77	11.18	13.07	58.81
C_OS	49.17	36.44	−5.54	36.86	351.35
C_OS_EL	43.73	38.36	−5.45	38.75	351.91
C_OS_Fe	42.28	18.51	−7.48	19.96	338.00
C_OS_Fe_EL	39.77	20.68	−8.20	22.24	338.38

Table 4. Color parameters of cochineal dyed cotton fabrics after one washing cycle.

Sample	L*	a*	b*	C*	h°
C_W	94.32	0.56	1.96	2.04	73.95
C_EL_W	92.50	2.20	0.75	2.33	18.85
C_Fe_W	81.29	6.44	15.88	17.14	67.93
C_Fe_EL_W	81.96	5.89	13.83	15.03	66.94
C_OS_W	51.93	34.89	−5.46	35.32	351.11
C_OS_EL_W	45.26	37.44	−5.50	37.84	351.65
C_OS_Fe_W	46.33	18.02	−6.48	19.15	340.23
C_OS_Fe_EL_W	42.97	20.47	−7.56	21.82	339.73

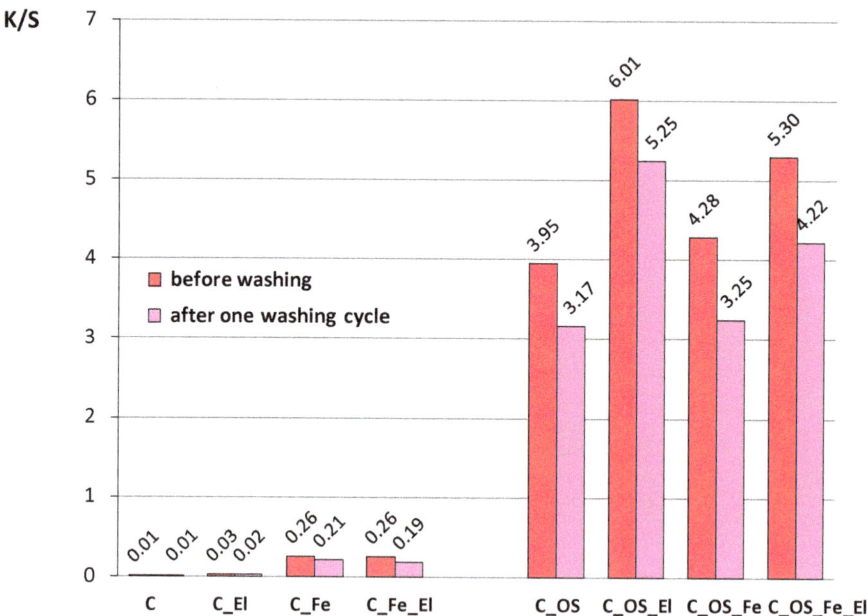

Figure 5. Color depth (K/S) of dyed cotton fabrics before and after one washing cycle (at 520 nm).

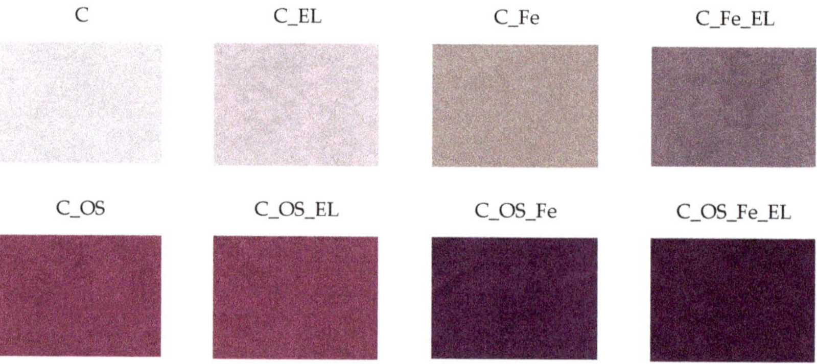

Figure 6. Cotton fabrics dyed with natural cochineal dye.

The results of color parameters presented in Tables 3 and 4 show a color analysis of cotton fabrics pre-treated with different treatments dyed with cochineal dye. Bleached cotton fabric dyed with cochineal dye had a lightness of 94.20 and chroma of 2.19, which correspond to low exhaustion from the dye bath. By adding electrolytes, exhaustion was higher than what resulted in the lightness decreasing and the chroma increasing (C*), but the change was not significant. Pre-mordanting with iron(II) sulfate heptahydrate further improved the absorption of the dye, which led to a decrease in lightness (L*), increased chroma, and a change in color, which is the effect of the binding of iron to the fabric. The C_Fe_EL sample took on a slightly different color, which can be seen in Figure 6. Cationization led to significantly higher exhaustion of dye, which was manifest by a different color hue, a decrease in lightness to 49.17, and an increase in chroma to 36.86. The color hue is in between red and purple. Pre-mordanting of cationized cotton with iron(II) sulfate heptahydrate led to an even greater reduction in lightness and the appearance of a purple color hue. The sample C_OS_EL had the highest chroma, which previously showed the highest exhaustion of the dye from the bath. After the first washing cycle, all samples had a slight increase in lightness, which means that some particles of dye were washed out from the surface. However, it was still significantly higher than without any pre-treatment.

Color depth (K/S) at 520 nm, presented in Figure 5, shows correlation with chroma and dye exhaustion. It is also visible by observation of dye samples (Figure 6). The K/S values obtained very clearly confirm the positive effect of cotton cationization on the natural dye exhaustion and the color depth achieved. Bleached cotton fabric had the smallest K/S value and sample C_OS_EL had the highest. It is visible that cationized cotton fabric has better K/S than noncationized cotton fabrics. By adding electrolytes, K/S changes to a higher level. The reason for this is the change of surface charge in cationization. Except for hydroxyl and carboxyl groups, cationized cotton contains amino groups as well. Therefore, when immersed in water, the amino and carboxyl groups exist in the ionized or zwitterion form.

By adding salt, the diffusion of dye is enhanced. The smallest and most rapidly diffusing ions are quickly adsorbed while the larger and more slowly diffusing dye anions follow, and the kinetics of binding cochineal dye is similar to the one with wool [24,59].

Color levelness describes the uniformity of color in different locations of the fabric. According to visual assessment, supported by Figure 6, it can be seen that cotton fabrics dyed with natural cochineal dye have good levelness regardless of pre-treatment. Small differences in color can be detected by human sight; however, it is quite difficult to quantify color differences. For that reason exactly, spectral remission was measured on random locations "until tolerance". The observed levelness, especially of cationized cotton fabrics, which exhaust 30–50% of natural dye, confirms that the cationic compound is evenly distributed and trapped between the cellulose chains, so the dyeing process is uniform as well.

Note that the observed phenomena, or K/S values, do not change after one washing cycle. Color fastness analysis was performed by calculating the total color difference (ΔE_{CMC}) between unwashed and washed dyed cotton fabrics. It was calculated after the first washing cycle and presented in Figure 7. From the results obtained, it is visible that the change in color occurred after the first washing cycle, but values of color differences (ΔE_{CMC}) are within the tolerance limits ($\Delta E_{CMC} \leq 2$) for cotton fabrics that were not pre-mordanted. The reason can be in yellowing after the pre-mordanting process, which significantly changed whiteness degree and therefore results in a changed hue, for the difference when Fe salt was not applied. The smallest ΔE_{CMC} is on the cotton fabric due to the lowest dye exhaustion. For the fabrics that had high exhaustion the total color difference is higher. Those fabrics exhaust five times more dyestuff, so it is logical that all exhausted natural dye could not be bonded to fibers, and subsequently, it was washed from the fabric and resulted in a higher color difference. It is important to emphasize that the color achieved after the first washing cycle on cationized fabrics was still five times higher than when not cationized.

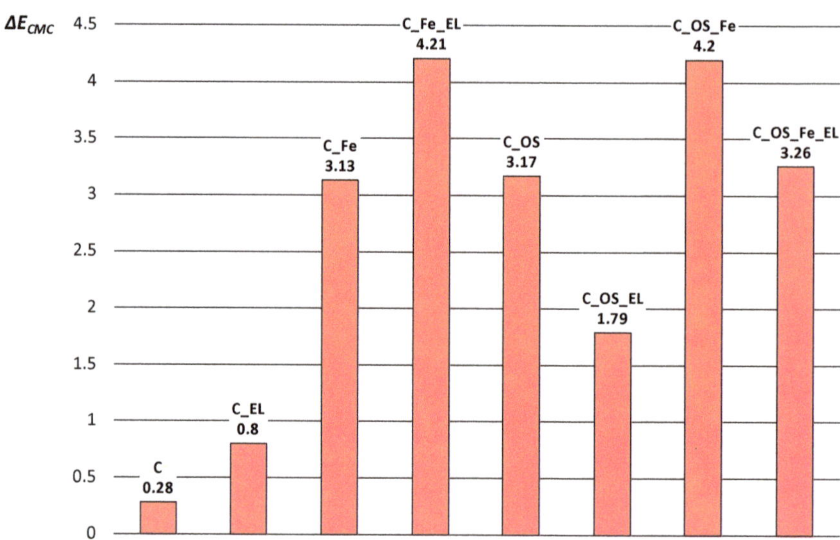

Figure 7. Color fastness of treated fabrics.

4. Conclusions

Cationized cotton fabrics adsorb more natural cochineal dye than the noncationized fabrics. When cationized, cotton possesses amino groups together with hydroxyl and carboxyl ones, so the adsorption of anionic dye is very rapid due to attraction. Electrolyte addition contributes to better diffusion of the dye. The best dyeing effect was achieved with cationization and electrolyte addition—the highest absorption rate, the highest chroma, and an acceptable color fastness ΔE_{CMC}. The results achieved with and without electrolyte are quite similar, so an electrolyte is not necessary. In the case of noncationized fabrics, an addition of electrolyte is important for exhaustion. Pre-mordanting with Fe-salt increases dye exhaustion and color depth, but not much as the cationization process. Pre-mordanting and cationization showed synergism considering dye exhaustion. The exhaustion is higher when the electrolyte was used, but from chromacity achieved and visual assessment it is not necessary. If only cationization is performed, the pure purple hue of cochineal is achieved. If violet hue needs to be achieved, mordanting with Fe-salt is necessary. Thus, when the process of cationization is performed before dyeing with natural dye, salt can be reduced or even unnecessary, making it friendlier for the environment.

Author Contributions: I.Č. and A.T. performed modification; I.Č. and A.T. performed electrokinetic analysis; I.Č., I.B. and A.S. performed pre-mordanting and dyeing; I.Č. and A.S. performed spectral analysis of dye and fabrics; A.S. and A.T. designed the study. All authors were included in writing and preparing the original draft. All authors have read and agreed to the published version of the manuscript.

Funding: This work has been supported in part by the Croatian Science Foundation under the project UIP-2017-05-8780 HPROTEX and in part by the University of Zagreb (TP6/21 and TP19/21).

Institutional Review Board Statement: Not applicable.

Informed Consent Statement: Not applicable.

Data Availability Statement: Data are available in a publicly accessible repository.

Acknowledgments: The work of doctoral student Ivana Čorak has been supported in part by the "Young researchers" career development project—training of doctoral students" of the Croatian Science Foundation. Any opinions, findings, and conclusions or recommendations expressed in this material are those of the authors and do not necessarily reflect the views of Croatian Science

Foundation. A paper recommended by the 14th Scientific and Professional Symposium Textile Science and Economy, The University of Zagreb Faculty of Textile Technology.

Conflicts of Interest: The authors declare no conflict of interest.

Sample Availability: Samples of the compounds are available from the authors.

References

1. Arora, J.; Agarwal, P.; Gupta, G. Rainbow of Natural Dyes on Textiles Using Plants Extracts: Sustainable and Eco-Friendly Processes. *Green Sustain. Chem.* **2017**, *7*, 35–47. [CrossRef]
2. Zarkogianni, M.; Mikropoulou, E.; Varella, E.; Tsatsaroni, E. Colour and fastness of natural dyes: Revival of traditional dyeing techniques. *Color. Technol.* **2010**, *127*, 18–27. [CrossRef]
3. Deveoglu, O. A Review on Cochineal (*Dactylopius Coccus Costa*) Dye. *Res. J. Recent Sci.* **2020**, *9*, 37–43.
4. Rehman, F.; Sanbhal, N.; Naveed, T.; Farooq, A.; Wang, Y.; Wei, W. Antibacterial performance of Tencel fabric dyed with pomegranate peel extracted via ultrasonic method. *Cellulose* **2018**, *25*, 4251–4260. [CrossRef]
5. Maxia, A.; Meli, F.; Gaviano, C.; Picciau, R.; De Martis, B.; Kasture, S.; Kasture, V. Dye plants: Natural resources from traditional botanical knowledge of Sardinia Island, Italy. *Indian J. Tradit. Knowl.* **2013**, *12*, 651–656.
6. Saakshy; Sharma, A.K.; Jain, R.K. Chapter in Biotechnology for Environmental Management and Resource Recovery. In *Application of Natural Dyes: An Emerging Environment-Friendly Solution to Handmade Paper Industry*; Kuhad, R.C., Singh, A., Eds.; Springer: New Delhi, India; Berlin/Heidelberg, Germany; New York, NY, USA; Dordrecht, The Netherlands; London, UK, 2013; pp. 279–288.
7. Parac-Osterman, Đ.; Karaman, B. *Osnove Teorije Bojenja Tekstila (Eng. Basics of Textile Dyeing Theory)*; University of Zagreb Faculty of Textile Technology: Zagreb, Croatia, 2013.
8. Sutlović, A. Study of Natural Dyestuff—Contribution to Human Ecology. Ph.D. Thesis, University of Zagreb Faculty of Textile Technology, Zagreb, Croatia, 2008.
9. Sutlović, A.; Parac-Osterman, Đ.; Đurašević, V. Croatian Traditional Herbal Dyes for Textile Dyeing. *TEDI* **2011**, *1*, 65–69.
10. Ansari, A.A.; Thakur, B.D. Extraction, characterization and application of a natural dye: The eco-friendly textile colorant. *Colourage* **2000**, *47*, 15–20.
11. Buttler Greenfield, A. *A Perfect Red, Empire, Espionage and the Quest for the Colour of Desire*; Harper Perennial: New York, NY, USA, 2005; pp. 103–117.
12. Brenko, A.; Randić, M. Exhibition *"The Power of Colour" (Moć boje)*; Etnografski Muzej: Zagreb, Croatia, 2009; pp. 7–153.
13. Borges, M.E.; Tejera, R.L.; Diaz, L.; Esparza, P.; Ibanez, E. Natural dyes extraction from cochineal (*Dactylopius coccus*). New extraction methods. *Food Chem.* **2012**, *132*, 1855–1860. [CrossRef]
14. Campana, M.G.; Robles Garcia, N.M.; Tuross, N. America's red gold: Multiple lineages of cultivated cochineal in Mexico. *Ecol. Evol.* **2015**, *5*, 607. [CrossRef]
15. Canamares, M.V.; Garcia-Ramos, J.V.; Domingo, C.; Sanchez-Cortes, S. Surface-Enhanced Raman Scattering Study of the Anthraquinone Red Pigment Carminic Acid. *Vib. Spectrosc.* **2006**, *40*, 161–167. [CrossRef]
16. Dapson, R.W. The History, Chemistry and Modes of Action of Carmine and Related Dyes. *Biotech. Histochem.* **2007**, *82*, 173–187. [CrossRef] [PubMed]
17. Allevi, P.; Anastasia, M.; Bingham, S.; Ciuffreda, P.; Fiecchi, A.; Cighetti, G.; Muir, M.; Tyman, J. Synthesis of Carminic Acid, the Colourant Principle of Cochineal. *J. Chem. Soc. Perkin Trans. 1* **1998**, 575–582. [CrossRef]
18. Naturex Focuses on Clean & Clear Labels at IFT18 with Launch of Plant-Based Alternative to Edta. Available online: https://www.naturex.com/Media2/Press-releases/Naturex-focuses-on-clean-clear-labels-at-IFT18-with-launch-of-plant-based-alternative-to-EDTA (accessed on 27 December 2021).
19. Sutlović, A.; Brlek, I.; Ljubić, V.; Glogar, M.I. Optimization of Dyeing Process of Cotton Fabric with Cochineal Dye. *Fibers Polym.* **2020**, *21*, 555–563. [CrossRef]
20. Prikhodko, S.V.; Rambaldi, D.C.; King, A.; Burr, E.; Muros, V.; Kakoulli, I. New advancements in SERS dye detection using infrared SEM and Raman spectromicroscopy (μRS). *J. Raman Spectrosc.* **2015**, *46*, 632–635. [CrossRef]
21. Stathopoulou, K.; Valianou, L.; Skaltsounis, A.-L.; Karapanagiotis, I.; Magiatis, P. Structure elucidation and chromatographic identification of anthraquinone components of cochineal (*Dactylopius coccus*) detected in historical objects. *Anal. Chim. Acta* **2013**, *804*, 264–272. [CrossRef]
22. Hebeish, A.; Elnagar, K.; Shaaban, M.F. Innovative Approach for Effecting Improved Mordant Dyeing of Cotton Textiles. *Egypt. J. Chem.* **2015**, *58*, 415–430.
23. Arroyo-Figueroa, G.; Ruiz-Aguilar, G.M.L.; Cuevas-Rodriguez, G.; Gonzalez-Sanchez, G. Cotton fabric dyeing with cochineal extract: Influence of mordant concentration. *Color. Technol.* **2011**, *127*, 39–46. [CrossRef]
24. Valipour, P.; Ekrami, E.; Shams-Nateri, A. Colorimetric Properties of Wool Dyed with Cochineal: Effect of Dye-Bath pH. *Prog. Color Colorants Coat.* **2014**, *7*, 129–138.
25. Lokhande, H.T.; Dorugade, V.A.; Sandeep, R.N. Applicaton of Natural Dyes on Polyester. *Am. Dyest. Report.* **1998**, *87*, 40–50.
26. Samanta, A.K.; Singhee, D.; Sethia, M. Application of single and mixture of selected natural dyes on cotton fabric: A scientific approach. *Colourage* **2003**, *50*, 29–42.

27. Angelini, L.G.; Bertoli, A.; Rolandelli, S.; Pistelli, L. Agronomic potential of *Reseda luteola L.* as new crop for natural dyes in textiles production. *Ind. Crops Prod.* **2003**, *17*, 199–207. [CrossRef]
28. Bechtold, T.; Mahmud-Ali, A.; Mussak, R.A.M. Natural dyes for textile dyeing: A comparison of methods to assess the quality of Canadian golden rod plant material. *Dye. Pigment.* **2007**, *75*, 287–293. [CrossRef]
29. Bechtold, T.; Mahmud-Ali, A.; Mussak, R.A.M. Reuse of ash-tree (*Fraxinus excelsior* L.) bark as natural dyes for textile dyeing: Process conditions and process stability. *Color. Technol.* **2007**, *123*, 271–279. [CrossRef]
30. Micheal, M.N.; Tera, F.M.; Aboelanwar, S.A. Colour measurements and colourant estimation of natural red dyes on natural fabrics using differnt mordants. *Colourage* **2003**, *50*, 31–42.
31. Brückner, U.; Struckmeier, S.; Dittrich, J.H.; Reumann, R.D. Zur Echtheit von Färbungen mit ausgewählten Naturfarbstoffen auf Synthesefasergeweben. *Texilveredlung* **1997**, *32*, 112–116.
32. Sutlović, A.; Glogar, M.I.; Tarbuk, A. Cochineal Colored Cotton as UV Shield: UV Protective Properties of Cotton Material Dyed with Cochineal Dyestuff. In *Scientific Notes of the Color Society of Russia*; Griber, Y.A., Schindler, V.M., Eds.; Smolensk State University Press: Smolensk, Russia, 2020; pp. 71–78.
33. Haddar, W.; Ticha, M.B.; Guesmi, A.; Khoffi, F.; Durand, B. A novel approach for a natural dyeing process of cotton fabric with Hibiscus mutabilis (Gulzuba): Process development and optimization using statistical analysis. *J. Clean. Prod.* **2014**, *68*, 114–120. [CrossRef]
34. Haddar, W.; Elksibi, I.; Meksi, N.; Mhenni, M.F. Valorization of the leaves of fennel (Foeniculum vulgare) as natural dyes fixed on modified cotton: A dyeing process optimization based on a response surface methodology. *Ind. Crops Prod.* **2014**, *52*, 588–596. [CrossRef]
35. Ticha, M.B.; Haddar, W.; Meksi, N.; Guesmi, A.; Mhenni, M.F. Improving dyeability of modified cotton fabrics by the natural aqueous extract from red cabbage using ultrasonic energy. *Carbohydr. Polym.* **2016**, *154*, 287–295. [CrossRef]
36. Peran, J.; Ercegović Ražić, S.; Sutlović, A.; Ivanković, T.; Glogar, M.I. Oxygen Plasma Pre-Treatment Improves Dyeing and Antimicrobial Properties of Wool Fabric Dyed with Natural Extract from Pomegranate Peel. *Color. Technol.* **2020**, *136*, 177–187.
37. Haji, A. Plasma activation and chitosan attachment on cotton and wool for improvement of dyeability and fastness properties. *Pigment Resin Technol.* **2020**, *49*, 483–489. [CrossRef]
38. Haji, A. Improved natural dyeing of cotton by plasma treatment and chitosan coating; optimization by response surface methodology. *Cellul. Chem. Technol.* **2017**, *51*, 975–982.
39. Gulzar, T.; Adeel, S.; Hanif, I.; Rehman, F.; Hanif, R.; Zuber, M.; Akhtar, N. Eco-friendly dyeing of gamma ray induced cotton using natural quercetin extracted from Acacia bark (*A. Nilotica*). *J. Nat. Fibers* **2015**, *12*, 494–504. [CrossRef]
40. Tarbuk, A.; Sutlović, A.; Grancarić, A.M.; Kopanska, A.; Trela, N.; Draczyński, Z. The Modified Cotton Dyed with Juglans Regia L. without Mordant. In *Book of Proceedings of the 8th International Textile, Clothing & Design Conference—Magic World of Textiles*; Dragčević, Z., Hursa Šajatović, A., Vujasinović, E., Eds.; University of Zagreb Faculty of Textile Technology: Zagreb, Croatia, 2016; pp. 212–217.
41. Tarbuk, A.; Grancarić, A.M.; Leskovac, M. Novel cotton cellulose by cationisation during the mercerisation process—Part 1: Chemical and morphological changes. *Cellulose* **2014**, *21*, 2167–2179. [CrossRef]
42. Tarbuk, A.; Grancarić, A.M.; Leskovac, M. Novel cotton cellulose by cationisation during the mercerization—Part 2: Interface phenomena. *Cellulose* **2014**, *21*, 2089–2099. [CrossRef]
43. Rupin, M.; Veatue, J.; Balland, B. Utilization of reactive epoxy-ammonium quaternaries on cellulose treatment for dyeing with direct and reactive dyes. *Textilveredlung* **1970**, *5*, 829–838.
44. Lewis, D.M.; McIlroy, K.A. The Chemical Modification of Cellulosic fibers to Enhance Dyeability. *Rev. Prog. Color* **1997**, *27*, 5–17.
45. Hauser, P.J.; Tabba, A.H. Improving the Environmental and Economic Aspects of Dyeing Cotton. *Color. Technol.* **2001**, *117*, 282–288. [CrossRef]
46. Draper, S.L.; Beck, K.R.; Smith, C.B.; Hauser, P. Characterization of the Dyeing Behavior of Cationic Cotton with Acid Dyes. *AATCC Rev.* **2003**, *3*, 51–55.
47. Sutlović, A.; Glogar, M.I.; Čorak, I.; Tarbuk, A. Trichromatic Vat Dyeing of Cationized Cotton. *Materials* **2021**, *14*, 5731. [CrossRef]
48. Aktek, T.; Millat, A.K.M.M. Salt free dyeing of cotton fiber—A critical review. *Int. J. Text. Sci.* **2017**, *6*, 21–33.
49. Correia, J.; Rainert, K.T.; Oliveira, F.R.; Valle, R.C.S.C.; Valle, J.A.B. Cationization of cotton fiber: An integrated view of cationic agents, processes variables, properties, market and future prospects. *Cellulose* **2020**, *27*, 8527–8550. [CrossRef]
50. Choudhury, A.K.R. Coloration of Cationized Cellulosic Fibers—A Review. *AATCC J. Res.* **2014**, *1*, 11–19. [CrossRef]
51. Tarbuk, A. Interface Phenomena of Cationized Cotton. Ph.D. thesis, University of Zagreb Faculty of Textile Technology, Zagreb, Croatia, 2009.
52. Tarbuk, A.; Grancarić, A.M. Chapter 6 in Cellulose and Cellulose Derivatives: Synthesis, Modification and Applications, Part I: Cellulose Synthesis and Modification. In *Interface Phenomena of Cotton Cationized in Mercerization*; Mondal, I.H., Ed.; Nova Science Publishers: New York, NY, USA, 2015; pp. 103–126.
53. Grancarić, A.M.; Tarbuk, A.; Dekanić, T. Elektropozitivan pamuk (eng. Electropositive Cotton). *Tekstil* **2004**, *53*, 47–51.
54. Soljačić, I.; Žerdik, M. Cotton mercerization. *Tekstil* **1968**, *17*, 495–518.
55. Dinand, E.; Vignon, M.; Chanzy, H.; Heux, L. Mercerization of Primary Wall Cellulose and its Implication for the Conversion of Cellulose I to Cellulose II. *Cellulose* **2002**, *9*, 7–18. [CrossRef]

56. Stana-Kleinschek, K.; Strand, S.; Ribitsch, V. Surface Characterization and Adsorption Abilities of Cellulose Fibers. *Polym. Eng. Sci.* **1999**, *39*, 1412–1424. [CrossRef]
57. Grancarić, A.M.; Tarbuk, A.; Pušić, T. Electrokinetic Properties of Textile Fabrics. *Color. Technol.* **2005**, *121*, 221–227. [CrossRef]
58. Parac-Osterman, Đ. *Osnove o Boji i Sustavi Vrjednovanja, (Eng. Color Basics and Evaluation Systems)*; University of Zagreb Faculty of Textile Technology: Zagreb, Croatia, 2007.
59. Lewis, D.M. *Wool Dyeing*; Society of Dyers and Colourists: Bradford, UK, 1992.

Article

Evaluation of DNA-Damaging Effects Induced by Different Tanning Agents Used in the Processing of Natural Leather—Pilot Study on HepG2 Cell Line

Sanja Ercegović Ražić [1,*], Nevenka Kopjar [2], Vilena Kašuba [2], Zenun Skenderi [1], Jadranka Akalović [1] and Jasna Hrenović [3]

1 Department of Materials, Fibres and Textile Testing, University of Zagreb Faculty of Textile Technology, Prilaz Baruna Filipovića 28a, 10000 Zagreb, Croatia
2 Mutagenesis Unit, Institute for Medical Research and Occupational Health, Ksaverska Cesta 2, 10000 Zagreb, Croatia
3 Division of Biology, Faculty of Science, University of Zagreb, Rooseveltov Trg 6, 10000 Zagreb, Croatia
* Correspondence: sanja.ercegovic.razic@ttf.unizg.hr

Abstract: For a long time, the production and processing of cowhide was based on the use of chrome tanning. However, the growing problem with chromium waste and its negative impact on human health and the environment prompted the search for more environmentally friendly processes such as vegetable tanning or aldehyde tanning. In the present study, we investigated the DNA-damaging effects induced in HepG2 cells after 24 h exposure to leather samples (cut into 1×1 cm^2 rectangles) processed with different tanning agents. Our main objective was to determine which tanning procedure resulted in the highest DNA instability. The extent of treatment-induced DNA damage was determined using the alkaline comet assay. All tanning processes used in leather processing caused primary DNA damage in HepG2 cells compared to untreated cells. The effects measured in the exposed cells indicate that the leaching of potentially genotoxic chemicals from the same surface is variable and was highest after vegetable tanning, followed by synthetic tanning and chrome tanning. These results could be due to the complex composition of the vegetable and synthetic tanning agents. Despite all limitations, these preliminary results could be useful to gain a general insight into the genotoxic potential of the processes used in the processing of natural leather and to plan future experiments with more specific cell or tissue models.

Keywords: comet assay; %DNA in tail; genotoxicity; in vitro; tanning; natural leather

Citation: Ražić, S.E.; Kopjar, N.; Kašuba, V.; Skenderi, Z.; Akalović, J.; Hrenović, J. Evaluation of DNA-Damaging Effects Induced by Different Tanning Agents Used in the Processing of Natural Leather—Pilot Study on HepG2 Cell Line. *Molecules* **2022**, 27, 7030. https://doi.org/10.3390/molecules27207030

Academic Editor: Marek Kosmulski

Received: 17 August 2022
Accepted: 14 October 2022
Published: 18 October 2022

Publisher's Note: MDPI stays neutral with regard to jurisdictional claims in published maps and institutional affiliations.

1. Introduction

Natural leather is a unique biological material with a complex morphological structure, the processing of which is extremely complex and requires many process steps, the most important of which is tanning, through which its characteristic and unique physical, chemical and mechanical properties are obtained. The main components of natural leather are proteins—collagen, keratin, elastin and reticulin—which make up about 33% of leather. Collagen, as the main component of leather, is responsible for the strength and toughness of raw and finished leather. Keratin is the only protein found in hair and the second most abundant protein in leather. Elastin is found in the papillary layer and is removed when the leather is processed. Besides collagen, proteoglycans, hyaluronic acid and smaller amounts of carbohydrates are also present in the reticular layer. Lipids are present in both the papillary and the reticular layers, the concentration depending on the origin of the leather (e.g., bovine leather contains 2–6% lipids) [1,2]. Collagen fibrils form a complex network and their arrangement contributes significantly to the strength of collagen structures [3]. The fibrils are grouped into fibril bundles that form collagen fibres, and the fibres fuse together to form bundles of collagen fibres [4]. This hierarchical part of the collagen fibre is important because it depends on the opening of the fibre structure during preparation for the tanning process. The splitting of the fibre structure at this stage of processing is important to achieve the final softness and strength of the finished hide [5].

The finished hide is a product made by processing raw animal hide in the tanning process by various means, the most common of which is the chrome tanning process, which makes the hide more durable and supple, and prevents rotting. The tanning agents used in the tanning process react chemically with the collagen molecule that makes up the natural skin and stabilise the triple helix structure of the collagen nucleus, making the leather resistant to chemical, thermal and microbiological degradation. The tanning agents used are usually inorganic compounds such as chromium, aluminium, iron, titanium and organic compounds such as aldehydes and vegetable tannins, synthanes and combinations of these compounds [6]. It is very important that the tannins used have the affinity and ability to react with collagen and have the appropriate size to penetrate the collagen fibres and achieve the desired cross-linking of the structure [7].

The manufacture and processing of leather has been based mainly on the use of chrome tanning salts for the last two centuries. Chromium salts are involved in the processing of more than 90% of the leather produced worldwide [8]. However, the growing problem of chromium waste and its negative impact on human health and the environment is prompting a search for more environmentally friendly methods. Several alternative processes to chrome tanning, such as vegetable tanning or aldehyde tanning [9–11], have been developed to produce chrome-free leather. Although these processes are gradually gaining commercial importance, these materials often cannot match the properties of chrome-tanned leather.

Considering the above, in this study we wanted to assess how the tanning procedures used differ according to their toxic potential. As known, for each new material or newly developed technological process various safety issues must be known before it is put into widespread use. Otherwise, it is possible that the material could cause unwanted health effects in the end users.

Generally, the first step in the evaluation of potential toxic effects of unknown or poorly known mixtures of chemicals is selection of a suitable test model. Testing strategies often are complex; they should be performed in some logical sequence, and usually start with the in vitro assays. In further steps, depending on the preliminary results, investigations extend to the higher levels of biological organisation, where various in vivo models are used.

Nowadays, in vitro testing relies on the use of different cell lines. Each of them has its own specific characteristics, according to which the best model for testing and evaluating the effects could be selected. For the purpose of the present study, we selected the HepG2 cell model, which has already proven useful for studying the genotoxicity of many direct and indirect mutagens and compounds with unknown or poorly known mechanisms of action [12,13]. As HepG2 cells were successfully used in several previous studies which investigated toxic potential of various dyes and agents common in the textile and leather industry [14–19], it seems reasonable to select them for this experiment. We considered HepG2 cells a suitable test system because they retain—to a certain extent—the activity of metabolic enzymes that are important for the biotransformation of chemicals in the liver, and exhibit many of the genotypic and phenotypic characteristics of liver cells [20–22]. Keeping these facts in mind, for the purposes of the pilot experiment we first selected that specific cell model. The logic of such a choice was to assess whether there will be measurable DNA damaging effects on the HepG2 cell model, and in accordance with the obtained results, to propose directions for further research. This research was intended as a preliminary assessment of the DNA-damaging potency of different procedures in leather processing and their possible harmful effects on DNA integrity. We investigated the DNA-damaging effects induced in HepG2 cells after 24 h of exposure to leather samples processed with different tanning procedures. This research starts from a hypothesis that different tanning procedures could result in varying degrees of DNA damage. In line with this, our main objective was to determine which tanning process resulted in the highest

DNA instability in the model cells. For this reason, the alkaline comet assay (single cell gel electrophoresis) was applied, which is considered one of the elementary tests for screening and early assessment of DNA-damaging effects. It enables detection of primary DNA damage at the level of an individual cell, inflicted by direct action of various chemical (or physical) agents or by indirect action of free radicals (for example, reactive oxygen species). The method was named after the appearance of the pattern of damaged DNA, which, stained with specific dyes and observed under a fluorescence microscope, resembles a celestial body—"comet". The method constists of several steps, starting with embedding of cells in agarose microgel, lysis of their cytoplasm and membrane structures, denaturation with alkaline buffer—which potentiates removal of histones—strand separation and release of relaxed DNA loops. During electrophoresis, DNA loops migrated from the comet head towards the anode. Their migration pattern can be visualised after staining with fluorescent dyes that bind to the DNA. The measurement of comets can be performed under a fluorescence microscope using image analysis software or by manual scoring. The most important descriptors of DNA damage are "tail intensity" (the percentage of DNA that has migrated into the comet tail) and "tail length" (the length of DNA migration expressed in μm). The measured data are automatically stored in the form Microsoft Excel sheets, and what follows is further mathematical and statistical processing. The alkaline comet assay identifies a broad spectrum of lesions: Single and double-strand breaks in DNA, alkali-labile sites, single-strand breaks associated with incomplete excision repair, DNA–DNA or DNA–protein cross-links [23–26].

Based on all previously mentioned facts, we anticipate that the results obtained using the proposed in vitro experimental design could contribute new basic knowledge useful for the safety assessment of the studied tanning agents, which is important to identify potentially harmful tanning processes and consequently ensure the safety of leather products for consumers.

2. Results

After 24 h of incubation of HepG2 cells with leather samples, no changes in pH values were detected in the RPMI 1640 medium compared with the negative control sample.

The results of the alkaline comet assay (Figure 1) show statistically significant increases in the levels of %DNA in the tail (A) and tail length (B) in the treated HepG2 cells after 24 h of exposure to the tested leather samples compared to the negative control ($p < 0.05$; Mann–Whitney U test). The highest level of DNA damage was observed after exposure to the leather sample processed with vegetable tanning (label VEG-T), followed by the synthetic tanning (label SYN-T), and two chrome tanning procedures (labels CHR-T1 and CHR-T2). The levels of both comet descriptors determined in the positive control were significantly increased compared to all other experimental groups.

Typical appearances of the comets corresponding to different degrees of DNA damage are shown in Figure 2.

Further evaluations of DNA damage included analysis of the frequency distribution of comets measured in treated and control cells (Figure 3). The Y-axis represents the number of comets belonging to a particular category in relation to four quartiles: <25th percentile, 25th–50th percentile, 50th–75th percentile and >75th percentile. The differences between the treated cells and the control were tested using Pearson's χ^2 test, and the statistically significant differences compared with the negative control are shown in the above figure. The pattern of DNA damage measured in HepG2 cells after exposure to the tested samples of leather shows that for both comet descriptors (%DNA in tail and tail length), the proportion of comets belonging to the <25th percentile decreased significantly compared to the untreated cells, while the proportion of those belonging to the 75th percentile decreased significantly. Although there were some minor variations in the percentages of comets belonging to a particular category with respect to the four quartiles, none of the differences observed after exposure to the leather sample obtained by chrome tanning (label CHR-T2) was statistically significant compared to the negative control.

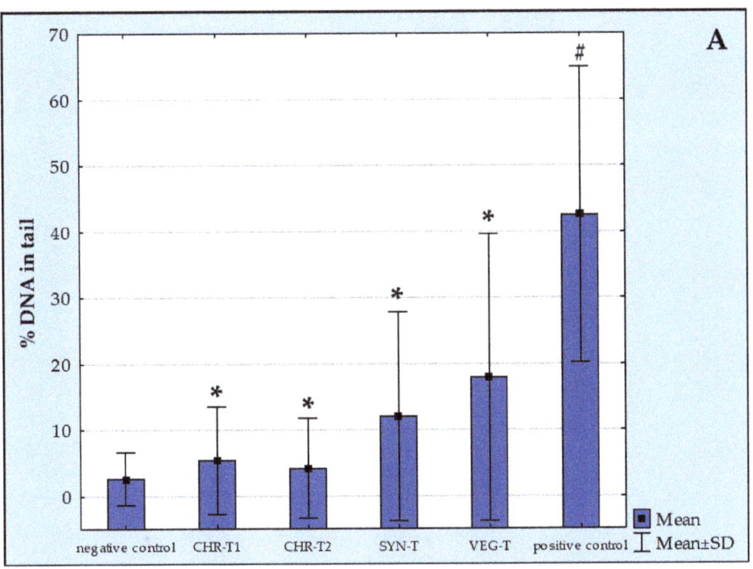

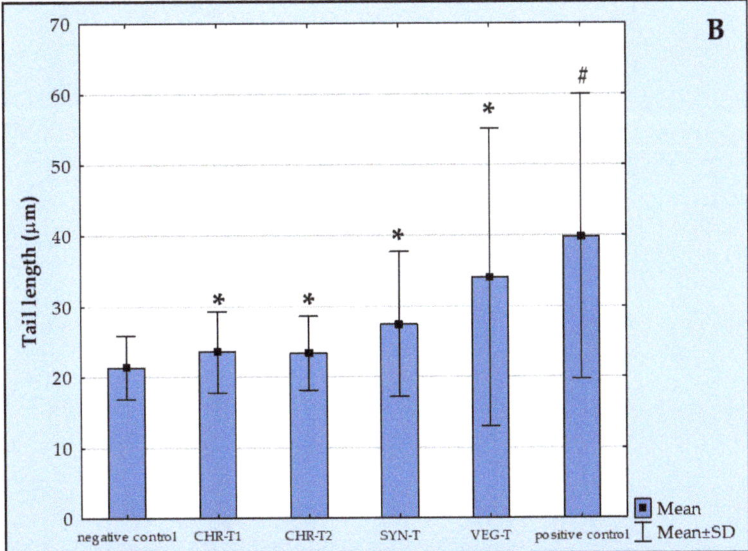

Figure 1. The extent of primary DNA damage in HepG2 cells measured by the alkaline comet assay after 24 h exposure to the leather samples tested and the corresponding negative and positive controls. The main comet descriptors were %DNA in tail (**A**) and tail length (**B**). CHR-T1 and CHR-T2 correspond to the leather sample processed with two chrome tanning procedures; SYN-T corresponds to the leather sample processed with synthetic tanning; VEG-T corresponds to the leather sample processed with vegetable tanning. Results are reported as mean ± standard deviations obtained by the measurements of 300 comets per experimental group. *—significantly increased compared to negative control; #—significantly increased compared to all other experimental groups ($p < 0.05$; Mann–Whitney U test).

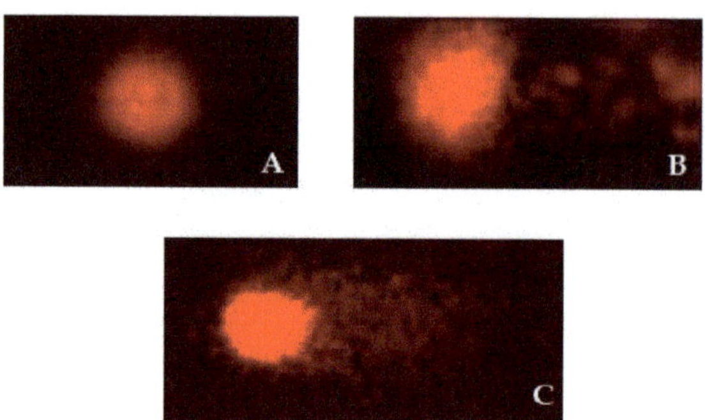

Figure 2. Photomicrographs show nucleoids of HepG2 cells observed in cell cultures after alkaline comet assay procedure: (**A**) Negative control without DNA damage; (**B**) positive control (hydrogen peroxide) with extensive DNA damage; (**C**) damaged DNA after 24 h exposure to leather sample processed with vegetable tanning. The photomicrographs were acquired using Comet Assay IV^{TM} image analysis software (Instem-Perceptive Instruments Ltd., Suffolk, Halstead, UK) at a magnification of 200×.

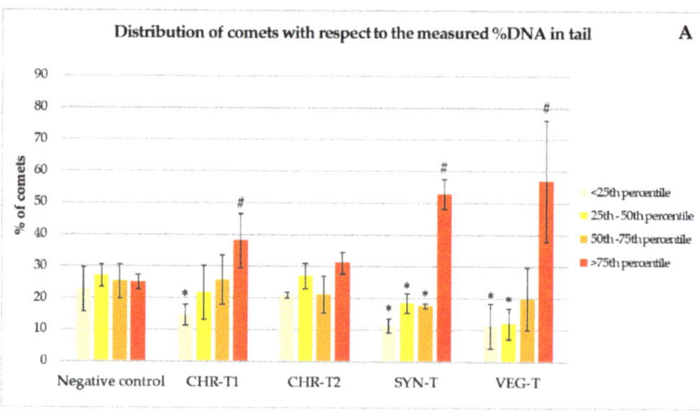

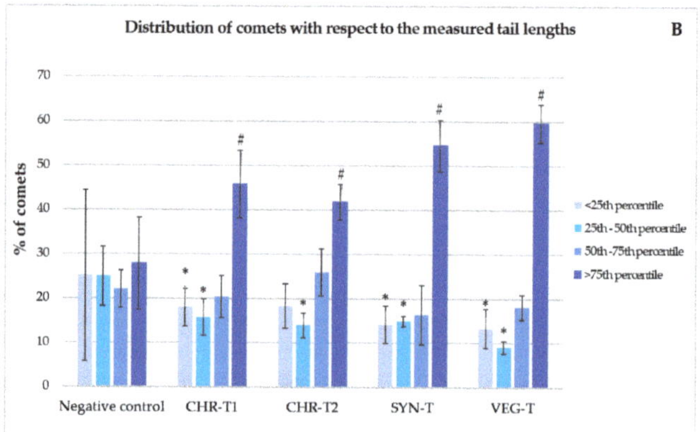

Figure 3. Frequency distribution of DNA damage (**A,B**) in HepG2 cells after 24 h exposure to the tested leather samples and the corresponding negative control. CHR-T1 and CHR-T2 correspond to the leather sample processed with two chrome tanning procedures; SYN-T corresponds to the leather sample processed with synthetic tanning; VEG-T corresponds to the leather sample processed with vegetable tanning. Results are reported as mean ± standard deviations of three independent evaluations that each separately included measurement of 100 comets. *—significantly decreased ($p < 0.05$) vs. negative control; #—significantly increased ($p < 0.05$) vs. negative control (Pearson's χ^2 test).

3. Discussion

The results of this pilot study confirmed that all tanning methods used in leather processing lead to primary DNA damage in HepG2 cells measurable by the alkaline comet assay. We also documented that "standard" chrome tanning, synthetic tanning and vegetable tanning differed in their ability to produce DNA lesions in HepG2 cells. Before going into detail about the significance of our results, we must point out that the available literature does not contain information about potentially DNA-damaging effects of tanning processes, nor does it provide recommendations on how to test them for their potentially harmful effects. Therefore, we chose the HepG2 cell model for this pilot study, which is considered useful for investigating the genotoxicity of compounds with unknown or poorly known mechanisms of action [20,27].

This research represents the first study that applied such an experimental design with the HepG2 cell model and tanning agents. In the light of real application, such basic in vitro research cannot be directly applied. We are very aware of the limitations and methodological constraints of the model used. We also anticipated potential problems in extrapolating the obtained results to the real exposure scenario and did not intend to assume analogous results in humans. Although results of in vitro studies do not provide definite answers on potential toxic outcomes following the application of some procedure, they could point to the risks associated with exposure to the tested agents which were observed at the lowest, i.e., cellular, level of the biological organisation. Their value is also reflected in the fact that every positive result is basically a warning sign that several other additional tests must be carried out to avoid any risk for the end users of materials processed using potentially harmful procedures. Our experimental design included a 24 h in vitro exposure of HepG2 cells to the leather samples tested. This period was long enough to cause the release of various compounds used for tanning and dyeing from the tested materials into the liquid cell medium RPMI 1640. This was demonstrated by observing visual colour changes of the medium. However, this was not accompanied by pH imbalance, as confirmed by measurements of the pH of the medium after treatment. Unfortunately, because this pilot study did not involve complex chemical analyses, specific detection of chemicals released into the medium or determination of their total amounts, we cannot determine exactly which specific compounds were leached from a particular leather sample. We can only relate the measured genotoxic effects to the known composition of the tanning and dyeing chemicals used in the processing of the leather samples used in this study and make assumptions as to what may have contributed to the results obtained.

To assess the genotoxic effects of chemicals released from leather samples processed by different methods, we incubated HepG2 cells with the leather rectangles cut into equal areas (1×1 cm^2). The effects measured in the exposed cells show that the leaching of potentially genotoxic chemicals from the same area is different; it was highest after vegetable tanning, followed by synthetic tanning and chrome tanning. The fact that vegetable tanning causes high levels of DNA damage in HepG2 cells is disappointing, considering that such leather processing methods are considered more harmless and environmentally friendly. However, from a genotoxicological point of view, the results obtained are not surprising, considering the conditions of cell exposure and the specificity of the model used. Leaching is influenced by the solubility of the chemicals used in the tanning (and dyeing) processes. Since the RPMI 1640 medium used to culture HepG2 cells is a liquid whose properties allow mobilisation of hydrophilic compounds [28], it is possible that many of the components used for vegetable tanning are readily released into the medium after 24 h of incubation due to their chemical properties and also enter the HepG2 cells and cause detectable levels of primary DNA damage. A high level of primary DNA damage measured after 24 h does not only mean a high genotoxic risk. Rather, it is a warning sign indicating that chemicals released from a tested sample have a high potency to produce DNA lesions that can be specifically detected by the alkaline comet assay. However, the genotoxic effects are much more complex and depend not only on the overall extent of primary DNA damage measured, but also on the efficiency of its repair and the persistence of unrepaired lesions.

The majority of lesions detected by the alkaline comet assay are single-strand breaks and alkali-labile sites in DNA [24,26,29], which are usually extensively repaired and do not cause significant damage. Other lesions detectable with the comet assay include single-strand breaks associated with incomplete excision repair, double-strand breaks, DNA–DNA and DNA–protein crosslinks [23,24]. Some of the DNA damage detectable by the assay is also due to DNA repair processes causing additional lesions [25].

So, what happened after the HepG2 cells came into contact with vegetable-tanned leather? The increased levels of primary DNA damage measured in these cells were apparently due to complex interactions between highly reactive compounds of natural origin that were efficiently released into the cell medium due to their high solubility. Looking at the composition of the substances used for vegetable tanning, it becomes clear that it is a very complex mixture of potentially reactive chemicals. Despite the widespread assumption that natural compounds generally have better biocompatibility and are less harmful to cells, tissues and organisms, it must be emphasised that many chemicals of natural origin in fact express their dual nature and act as both prooxidants and antioxidants, depending on the concentration [30,31]. As is well known, prooxidant behaviour depends largely on the environment in which these compounds are present. It is also modulated by the presence of O_2 and some redox-active metal ions such as iron and copper, which can lead to oxidative DNA degradation [31,32]. Therefore, chemicals of natural origin used for vegetable tanning could also be genotoxic, depending on their concentration. Tannins are known to be natural products found in most higher plants. They are usually divided into water-soluble (hydrolysable) polyphenols, such as gallo-tannins and ellagi-tannins, and condensed polyflavonoid tannins, which are rarely hydrolysed [33]. The vegetable tanning process used in the processing of a leather sample examined in the present study used tannins from mimosa, chestnut and quebracho. These are now considered the most important sources of plant tannins for leather production, but also represent industrially available natural sources of polyphenols used for many other purposes [34–36]. It has long been known that tannin-related substances induce DNA fragmentation [37]. Therefore, the results of the comet assay performed in our study are highly justified by the existing literature. Tannic acid is one of the most important hydrolysable tannins [33]. As early as the mid-1990s, Bhat and Hadi [38,39] observed that tannic acid causes DNA degradation in the presence of Cu (II) through the formation of reactive oxygen species. The cytotoxic and DNA-damaging effects of this compound were recently confirmed using the same cell model as ours, HepG2 cells. Mhlanga et al. [40] found that tannic acid increased both cell death and DNA fragmentation. These results are consistent with our own observations. Nowadays, synthetic tannins are increasingly used and research in this field is growing. Uddin et al. [41] recently investigated the role of glutaraldehyde in the tanning process compared to conventional chrome tanning. Their results suggest the usefulness of glutaraldehyde, especially in terms of minimising chromium pollution and reducing the generation of toxic waste and its impact on the environment. Our results regarding the genotoxicity of leather processed with synthetic tannins are not very convincing—we observed relatively high DNA damage in HepG2 cells exposed to this tested material. The leather sample obtained with synthetic tanning that we tested in the present experiment was processed with the agents Sellatan P and Sellatan CF based on modified polysulphonic acids and glutaraldehyde. Available literature reports indicate that these compounds can cause primary DNA damage. The main problem with glutaraldehyde is that this compound causes DNA–protein cross-linking [42–44]. It is important to note that repair of these lesions leads to DNA excision [42], resulting in additional "repair-related" primary DNA damage that can be detected by the alkaline comet assay. Taken together, previously known mechanisms of DNA damage induced by this specific chemical and the results we obtained in the present study match very well. Here, we must emphasise that the sample of leather processed with the synthetic tanning was used in our research only for the purpose of comparison with other samples. The aim of our research was not to collect evidence in support of this type of leather tanning, especially considering that in the meantime, glutaraldehyde was listed on the Substances of Very High Concern (SVHC) candidate list by the European Chemicals Agency [45]. However, the scientific project of which part is the research described in the present manuscript has been conducted continuously over several years, when the ban of glutaraldehyde was not yet in force. As far as we know, the ban happened in 2021, while our research was conducted mostly prior, or parallel to that ban. This means that at the time when our experiments were conducted, we could not predict that the ban would come into force. It is clear to us that prohibited substances may no longer be used in processing of leather, but for the purpose of experiments it is allowed to compare the effects of prohibited substances with those whose use is permitted. More than 90% of the leather produced worldwide is tanned with chromium [8]. However, doubts remain about the amount of chromium released from leather and its impact on human health and the environment. EU legislation states that "leather products that come into direct, prolonged or repeated contact with the skin shall not be placed on the market if the leather contains chromium (VI) in concentrations equal to or greater than 3 mg/kg" [46]. Thus, if the level of toxic Cr in various goods, garments and footwear is kept very low, the risk to consumers could be acceptable. Our results suggest that chromium tanning leads to slightly less primary DNA damage in HepG2 cells compared to tanning with synthetic and vegetable tanning agents. From a toxicological point of view, this is not surprising, considering that the chromium (III) used for tanning is a less hazardous species than chromium (VI) [47,48]. We assume that under our specific experimental conditions, the total amount of chromium (III) leached from chrome-tanned leather was relatively low compared to the amounts of other reactive substances released from the same surface of the other leather samples. Based on the results obtained, it is possible that the cell exposure conditions did not promote the conversion of the released Cr (III) to the more toxic Cr (VI), which is known to be a strong oxidant in acidic media [49]. However, this does not seem to have been the case, which was also supported by the finding that the pH of the medium did not differ from that of the negative control culture. We also found that dyeing chrome-tanned leather with black dye did not cause additional DNA damage compared to chrome-tanned leather that was not further processed. This result could be related to the negligible leaching of the dye into the culture medium, which was documented by the fact that the colouration of the medium did not change noticeably after 24 h of incubation with the sample of black-dyed leather. As this study was a pilot study with only one experimental scenario, we cannot answer whether longer exposure times could lead to greater leaching of chromium or dye and what effects could be expected under these conditions. This needs to be investigated in future studies with a wider range of exposure times. We should also briefly point out the limitations of the experimental design used. Although the model system used here was suitable for a general assessment of DNA-damaging effects, it unfortunately cannot reproduce real-life exposure well. In fact, no in vitro cell system, not even one using skin cells, can authentically mimic exposure from skin contact with a particular material. This is because the cells are grown in a liquid medium, unlike skin cells, which are organised in a very specific tissue. Leather products are usually in close contact with the surface of human skin, which has specific properties and multiple protective barriers against potentially harmful exposures. Such complex protective barriers are absent in cells maintained in cultures, which are more vulnerable and susceptible to higher levels of DNA damage. In addition, the leaching of potentially harmful substances from the skin surface after direct contact with any material is greatly reduced compared to in vitro conditions because the skin surface is normally relatively dry and the intercellular pores are largely impermeable to chemicals. This is the main difference between the exposure of the cells used in our experimental conditions and the real exposure of skin cells. In addition, the risk of harmful exposure following direct contact with a material containing potentially toxic substances would be greater for hydrophobic chemicals, as the outer stratum corneum of the skin is lipophilic. Therefore, under real conditions, one might expect a slower release of the hydrophilic chemicals than was the case under our exposure conditions. With this experimental model, we also cannot predict the extent to which potentially harmful chemicals will penetrate into the deeper layers of the skin and potentially cause significant toxic effects there.

4. Materials and Methods

4.1. Leather Samples

The samples of natural (bovine) leather used in this study were processed according to the same procedures before tanning and dyeing. Three samples were semi-processed bovine leathers (labelled CHR-T1, SYN-T and VEG-T) tanned with different tanning agents (chrome, synthetic and vegetable tanning agents), and one black-dyed leather sample (CHR-T2), which was dyed after the tanning process and represented the final product. All tested samples were produced by a Croatian company (PSUNJ factory, Rešetari, Croatia). A detailed description of the different tanning agents used in leather production can be found in Table 1.

Table 1. Basic parameters of the technological processes in the tanning of leather with different tanning agents.

Labelling of Leather Samples	CHR-T1	CHR-T2	SYN-T	VEG-T
Pickling process	1.6–1.8% acid (formic acid, sulphuric acid) 5–7% sodium chloride 40–45% water temp. 19–22 °C pH 3.00	1.6–1.8% acid (formic acid, sulphuric acid) 5–7% sodium chloride 40–45% water temp. 19–22 °C pH 3.00	2.0–2.5% commercial product based on polysulphonic acid, without salts 50% water temp. 20–25 °C	/
Pre-Tanning process	/	/	/	7–9% synthetic tanning agent for better tanning process; prevents the reduction of the concentration of the plant extracts) 20% water temp. 20–23 °C pH 5.0–5.5
Scouring	/	/	/	water, temp. 30 °C, circular bath
Tanning process *	3.2–3.6% basic chromium sulphate (commercially agent 25–27% Cr_2O_3; 33⁰Sch) pH 2.4	3.2–3.6% basic chromium sulphate (commercially agent 25–27% Cr_2O_3; 330Sch) pH 2.4	1.5–2.0% synthetic tanning agent based on aliphatic polyaldehyde, metal free 2.0–2.5% commercial product based on polysulphonic acid, 0.1–0.2% formic acid (85%), pH 3.5 0.1–0.2% sodium bicarbonate 0.1–0.2% sodium bisulphite, pH 3.8–4.0	4–6% mimosa; 9–11% chestnut; 9–11% quebracho 1.5–2.5% synthetic tanning agent (for softness, suppleness and strength, dyeing) water 20% pH 3.0–3.5
Basification process	0.25–0.32% basifying agents (commercial preparations of salt mixtures with low alkaline reactivity)—pH 10–12), water dispersed fungicide with zero volatile organic compounds; 20–25% water temp. 50 °C	0.25–0.32% basifying agents (commercial preparations of salt mixtures with low alkaline reactivity)—pH 10–12), water dispersed fungicide with zero volatile organic compounds; 20–25% water temp. 50 °C	/	/
Dyeing	/	Aniline Black dyes	/	/

Note: The quantity of tanning agents and other agents used to process the leather is not constant and is always adjusted depending on the quality of the raw or semi-finished product that is processed into finished leather, the quality of which must meet the buyer's criteria and is the responsibility of the manufacturer. * Depending on the tanning agent, the technological processes differ in terms of concentration, treatment time, pH, water content and chemical auxiliaries in the tanning bath required to obtain leather with high quality and good properties.

4.2. Cell Culture

The human hepatoma cell line HepG2 was obtained from the American Type Culture Collection (ATCC) (Manassas, VA, USA). Cells were cultivated in RPMI 1640 medium with antibiotics (penicillin/streptomycin) (Sigma-Aldrich, Steinheim, Germany) and 10% fetal bovine serum (FBS) (Sigma-Aldrich). They were seeded in 25 cm^2 flasks and maintained in a humidified atmosphere of 5% CO_2 at 37 °C (Heraeus Hera Cell 240 incubator, Langenselbold, Germany).

4.3. Experimental Design

Three independent experiments were performed with HepG2 cell cultures grown for 48 h before the treatments. Samples of leather (cut into rectangles of 1 × 1 cm^2) were first sterilised with UV light (30 min on each side of the leather sample) and then placed in the flasks containing the cell cultures in fresh complete medium. The cultures were incubated with the tested leather samples for 24 h in a humidified atmosphere of 5% CO_2 at 37 °C (Heraeus Hera Cell 240 incubator, Langenselbold, Germany). After incubation, the medium containing the leather samples was discarded and the pH was measured. The cells were washed with sterile PBS, trypsin-EDTA was added to each flask, and the flasks were further incubated at 37 °C. After detaching the cells, complete medium was added to inactivate trypsin; the cells were mixed gently with a pipette, transferred to tubes and centrifuged at 800× *g* for 4 min. The supernatant was drained off and small amounts of PBS buffer were added to the precipitate. The resulting cell suspensions were used to prepare agarose microgels.

4.4. Alkaline Comet Assay

To assess the extent of primary DNA damage in single cells, we used the alkaline comet assay procedure [50–52] with slight modifications [53]. Agarose microgels were prepared on slides pre-coated with 1% normal melting point (NMP) agarose. Duplicate slides were prepared for each experimental point. The first layer of the gel consisted of 0.6% NMP agarose. It was covered by: (1) A mixture of 0.5% low melting point agarose (LMP agarose) and cell samples (V = 15 μL suspension) and (2) the top layer of 0.5% LMP agarose. In our experiments, untreated cells served as negative controls. To prepare positive control slides, we exposed microgels containing untreated cells to 30 μmol/L hydrogen peroxide on ice for 10 min. Hydrogen peroxide was chosen for this purpose because it causes extensive DNA damage that can be detected by the method used. As a rule, the use of a known genotoxic substance as a positive control is recommended in order to obtain a positive reaction with the comet assay. After complete preparation, all slides were further processed in the same way. After polymerisation, the gels were immersed overnight in a lysis buffer (2.5 mol/L NaCl (Kemika, Zagreb, Croatia), 100 mmol/L Na$_2$EDTA, 10 mmol/L Tris-HCl, 1% sodium lauroyl sarcosinate, pH 10) containing 1% Triton X-100 and 10% dimethyl sulphoxide (Kemika, Zagreb, Croatia). The next day, slides were washed with distilled water to remove excess salt, immersed in freshly prepared cold denaturing/electrophoresis buffer (300 mmol/L NaOH, 1 mmol/L EDTA, pH > 13) and incubated in the dark at 4 °C for 40 min. The slides were then placed in a horizontal electrophoresis unit (SCIE PLAST, Cambridge, UK) filled with the same buffer. Electrophoresis took 20 min at 0.8 V/cm, 300 mA (power supply Power Pac HC^{TM}, BIO-RAD, Hercules, CA, USA) and 4 °C. The slides were then neutralised with three portions of Tris-HCl buffer (0.4 mol/L; pH 7.5). The microgels thus prepared were dehydrated with 70% ethanol for 10 min and with 96% ethanol for 10 min, air dried and stored in tightly closed plastic boxes protected from moisture and light until analysis. After staining with ethidium bromide (20 μg/mL), the microgels were analysed under an epifluorescence microscope (Olympus BX51, Tokyo, Japan) at 200× magnification. Analysis was performed by randomly selecting at least 50 comets per microgel using Comet Assay IV^{TM} software (Instem-Perceptive Instruments Ltd., Suffolk, Halstead, UK). As the experiment was performed in triplicate, a total of 300 individual comet measurements were taken for each experimental point. Two comet descriptors were selected to quantify the extent of DNA damage: %DNA in the tail and tail length (expressed in micrometres). Statistical calculations were performed using Statistica—Data Science Workbench software, version 14 (Licence No. 14.0.0.15; TIBCO Software Inc. 2020; Palo Alto, CA, USA). Basic descriptive statistics were calculated for each experimental point. Then, the Mann–Whitney U test was applied to make comparisons between the treated samples and the controls. Comparisons between the values obtained for the number of comets belonging to a given category with respect to four quartiles were made using Pearson's χ^2 test for two-by-two contingency tables [54]. The statistical significance level was set at $p < 0.05$.

5. Conclusions

Using the HepG2 cell model and the alkaline comet assay method, we have shown that all tanning methods used in leather processing can have genotoxic effects to some degree. Our experimental design allows a reliable estimation of the genotoxic effects mainly of the hydrophilic compounds present in the leather samples processed with different tanning methods. The results suggest that vegetable and synthetic tanning, perhaps due to their complex composition, cause higher overall primary DNA damage than "normal" chrome tanning.

Despite all limitations, we believe that the obtained results contribute interesting new information useful for the future safety assessments of the studied tanning agents. These preliminary results may be useful to gain a general insight into the genotoxic potential of the processes used in leather processing and to plan future experiments with more specific cell or tissue models. Future research regarding toxic aspects of vegetable tanning should focus on detailed phytochemical characterisation, explaining of interactions between the components included in the vegetable tanning procedures and determination of the concentration ranges in which their possible harmful effects would be minimised, while maintaining efficiency in leather processing. As the potentially useful cell model, skin cell lines are proposed, of which several commercially available types are available. Last, but not least, besides specific modifications of the comet assay, the use of cytogenetic assays is also advised. Among them, there are two in vitro tests proposed by the OECD: Chromosomal aberration test and micronucleus test. The latter test in its "cytome" version can provide a range of useful information valuable for further risk assessments.

Author Contributions: Conceptualisation, N.K., S.E.R. and Z.S.; methodology, N.K., S.E.R. and V.K.; software, N.K.; formal analysis, J.A., N.K., S.E.R. and V.K.; investigation, S.E.R. and Z.S.; resources, J.H.; data curation, N.K.; writing—original draft preparation, N.K., S.E.R. and V.K.; writing—review and editing, S.E.R., N.K. and J.A.; supervision, S.E.R. and Z.S.; project administration, Z.S.; funding acquisition, S.E.R., N.K., V.K. and Z.S. All authors have read and agreed to the published version of the manuscript.

Funding: This work has been fully supported by the Croatian Science Foundation under the project/grant number IP-2016-06-5278.

Institutional Review Board Statement: Not applicable.

Informed Consent Statement: Not applicable.

Data Availability Statement: Not applicable.

Conflicts of Interest: The authors declare no conflict of interest.

Sample Availability: Samples of the leathers are available from the authors.

References

1. Falkiewicz-Dulik, M.; Janda, K.; Wypych, G. *Handbook of Biodegradation, Biodeterioration and Biostabilization*, 2nd ed.; ChemTec Publishing: Toronto, ON, Canada, 2015; pp. 135–137.
2. Mark, H.F. *Encyclopedia Od Polymer Science and Technology*; John Wiley & Sons: Hoboken, NJ, USA, 2011.
3. Sizeland, K. Nanostructure and Physical Properties of Collagen Biomaterials. Ph.D. Thesis, Massey University, Manawatu, New Zealand, 2015.

4. Harris, S.; Veldmeijer, A.J. *Why Leather?* Sidestone Press: Leiden, The Netherlands, 2014.
5. Covington, T.; Wise, W.R. *Tanning Chemistry: The Science of Leather*, 2nd ed.; Royal Society of Chemistry: Cambridge, UK, 2019; pp. 1–7, 204–217, 318–324.
6. Krishnamoorthy, G.; Sadulla, S.; Sehgal, P.K.; Mandal, A.B. Green chemistry approaches to leather tanning process for making chrome-free leather by unnatural amino acids. *J. Hazard. Mater.* **2012**, *215–216*, 173–182. [CrossRef] [PubMed]
7. Lischuk, V.; Plavan, V.; Danilkovich, A. Transformation of the collagen structure during beam-house processes and combined tanning. *Proc. Estonian Acad. Sci.* **2006**, *12*, 188–198. [CrossRef]
8. Hedberg, Y.S.; Lidén, C.; Odnevall Wallinder, I. Chromium released from leather-I: Exposure conditions that govern the release of chromium(III) and chromium(VI). *Contact. Dermat.* **2015**, *72*, 206–215. [CrossRef] [PubMed]
9. Rolence, C.C.; Musabila, M.M.; Samwel, N.S.; Hilonga, A.; Kanth, S.V.; Njau, K.N. Alternative tanning technologies and their suitability in curbing environmental pollution from the leather industry: A comprehensive review. *Chemosphere* **2020**, *254*, 126804. [CrossRef]
10. Shirmohammadli, Y.; Efhamisisi, D.; Pizzi, A. Tanins as a sustainable raw material for green chemistry: A review. *Ind. Crops Prod.* **2018**, *126*, 316–332. [CrossRef]
11. Sathiyamoorthy, M.; Selvi, V.; Mekonneu, D.; Habtamu, S. Preparation of eco-friendly leather by process modifications to make pollution free tanneries. *J. Eng. Comput. Appl. Sci.* **2013**, *2*, 17–22.
12. Valentin-Severin, I.; Le Hegarat, L.; Lhuguenot, J.C.; Le Bon, A.M.; Chagnon, M.C. Use of HepG2 cell line for direct or indirect mutagens screening: Comparative investigation between comet and micronucleus assays. *Mutat. Res.* **2003**, *536*, 79–90. [CrossRef]
13. Fic, A.; Žegura, B.; Sollner Dolenc, M.; Filipič, M.; Peterlin Mašič, L. Mutagenicity and DNA damage of bisphenol A and its structural analogues in HepG2 cells. *Arh. Hig. Rada Toksikol.* **2013**, *64*, 189–200. [CrossRef]
14. Séverin, I.; Jondeau, A.; Dahbi, L.; Chagnon, M.C. 2,4-Diaminotoluene (2,4-DAT)-induced DNA damage, DNA repair and micronucleus formation in the human hepatoma cell line HepG2. *Toxicology* **2005**, *213*, 138–146. [CrossRef]
15. Tsuboy, M.S.; Angeli, J.P.; Mantovani, M.S.; Knasmüller, S.; Umbuzeiro, G.A.; Ribeiro, L.R. Genotoxic, mutagenic and cytotoxic effects of the commercial dye CI Disperse Blue 291 in the human hepatic cell line HepG2. *Toxicol. In Vitro* **2007**, *21*, 1650–1655. [CrossRef]
16. Oliveira, G.A.; Ferraz, E.R.; Chequer, F.M.; Grando, M.D.; Angeli, J.P.; Tsuboy, M.S.; Marcarini, J.C.; Mantovani, M.S.; Osugi, M.E.; Lizier, T.M.; et al. Chlorination treatment of aqueous samples reduces, but does not eliminate, the mutagenic effect of the azo dyes Disperse Red 1, Disperse Red 13 and Disperse Orange 1. *Mutat. Res.* **2010**, *703*, 200–208. [CrossRef] [PubMed]
17. Ching Chen, S.; Hseu, Y.C.; Sung, J.C.; Chen, C.H.; Chen, L.C.; Chung, K.T. Induction of DNA damage signaling genes in benzidine-treated HepG2 cells. *Environ. Mol. Mutagen.* **2011**, *52*, 664–672. [CrossRef] [PubMed]
18. Ferraz, E.R.; Umbuzeiro, G.A.; de-Almeida, G.; Caloto-Oliveira, A.; Chequer, F.M.; Zanoni, M.V.; Dorta, D.J.; Oliveira, D.P. Differential toxicity of Disperse Red 1 and Disperse Red 13 in the Ames test, HepG2 cytotoxicity assay, and Daphnia acute toxicity test. *Environ. Toxicol.* **2011**, *26*, 489–497. [CrossRef] [PubMed]
19. Ferraz, E.R.; Li, Z.; Boubriak, O.; de Oliveira, D.P. Hepatotoxicity assessment of the azo dyes disperse orange 1 (DO1), disperse red 1 (DR1) and disperse red 13 (DR13) in HEPG2 cells. *J. Toxicol. Environ. Health A* **2012**, *75*, 991–999. [CrossRef]
20. Knasmüller, S.; Parzefall, W.; Sanyal, R.; Ecker, S.; Schwab, C.; Uhl, M.; Mersch-Sundermann, V.; Williamson, G.; Hietsch, G.; Langer, T.; et al. Use of metabolically competent human hepatoma cells for the detection of mutagens and antimutagens. *Mutat. Res.* **1998**, *402*, 185–202. [CrossRef]
21. Westerink, W.M.; Schoonen, W.G. Cytochrome P450 enzyme levels in HepG2 cells and cryopreserved primary human hepatocytes and their induction in HepG2 cells. *Toxicol. In Vitro* **2007**, *21*, 1581–1591. [CrossRef]
22. Westerink, W.M.; Schoonen, W.G. Phase II enzyme levels in HepG2 cells and cryopreserved primary human hepatocytes and their induction in HepG2 cells. *Toxicol. In Vitro* **2007**, *21*, 1592–1602. [CrossRef]
23. Tice, R.R.; Agurell, E.; Anderson, D.; Burlinson, B.; Hartmann, A.; Kobayashi, H.; Miyamae, Y.; Rojas, E.; Ryu, J.C.; Sasaki, Y.F. Single cell gel/comet assay: Guidelines for in vitro and in vivo genetic toxicology testing. *Environ. Mol. Mutagen.* **2000**, *35*, 206–221. [CrossRef]
24. Collins, A.R. The comet assay for DNA damage and repair: Principles, applications, and limitations. *Mol. Biotechnol.* **2004**, *26*, 249–261. [CrossRef]
25. Azqueta, A.; Collins, A.R. The essential comet assay: A comprehensive guide to measuring DNA damage and repair. *Arch. Toxicol.* **2013**, *87*, 949–968. [CrossRef]
26. Langie, S.A.S.; Azqueta, A.; Collins, A.R. The comet assay: Past, present, and future. *Front. Genet.* **2015**, *6*, 1–3. [CrossRef] [PubMed]
27. Majer, B.J.; Mersch-Sundermann, V.; Darroudi, F.; Laky, B.; de Wit, K.; Knasmüller, S. Genotoxic effects of dietary and lifestyle related carcinogens in human derived hepatoma (HepG2, Hep3B) cells. *Mutat. Res. Mol. Mech. Mutagen.* **2004**, *551*, 153–166. [CrossRef] [PubMed]
28. Specification Sheet, Product Name: RPMI-1640 Medium with L-Glutamine and Sodium Bicarbonate, Liquid, Sterile-Filtered, Suitable for Cell Culture. Available online: https://www.sigmaaldrich.com/HR/en/specification-sheet/SIGMA/R8758 (accessed on 24 November 2021).

29. Møller, P.; Azqueta, A.; Boutet-Robinet, E.; Koppen, G.; Bonassi, S.; Milić, M.; Gajski, G.; Costa, S.; Teixeira, J.P.; Costa Pereira, C.; et al. Minimum Information for Reporting on the Comet Assay (MIRCA): Recommendations for describing comet assay procedures and results. *Nat. Protoc.* **2020**, *15*, 3817–3826. [CrossRef] [PubMed]

30. Martin-Cordero, C.; Leon-Gonzalez, J.A.; Calderon-Montano, J.M.; Burgos-Moron, E.; Lopez-Lazaro, M. Pro-oxidant natural products as anticancer agents. *Curr. Drug Targets* **2012**, *13*, 1006–1028. [CrossRef]

31. Carocho, M.; Ferreira, I.C.F.R. A review on antioxidants, prooxidants and related controversy: Natural and synthetic compounds, screening and analysis methodologies and future perspectives. *Food Chem. Toxicol.* **2013**, *51*, 15–25. [CrossRef]

32. Azmi, A.S.; Bhat, S.H.; Hanif, S.; Hadi, S.M. Plant polyphenols mobilize endogenous copper in human peripheral lymphocytes leading to oxidative DNA breakage: A putative mechanism for anticancer properties. *FEBS Lett.* **2006**, *580*, 533–538. [CrossRef]

33. Pizzi, A. Tannins: Prospectives and Actual Industrial Applications. *Biomolecules* **2019**, *9*, 344. [CrossRef]

34. Krisper, P.; Tisler, V.; Skubic, V.; Rupnik, I.; Kobal, S. The use of tannin from chestnut (*Castanea vesca*). *Basic Life Sci.* **1992**, *59*, 1013–1019. [CrossRef]

35. Missio, A.L.; Tischer, B.; dos Santos, P.S.B.; Codevilla, C.; de Menezes, C.R.; Barin, J.S.; Haselein, C.R.; Labidi, J.; Gatto, D.A.; Petutschnigg, A.; et al. Analytical characterization of purified mimosa (*Acacia mearnsii*) industrial tannin extract: Single and sequential fractionation. *Sep. Purif. Technol.* **2017**, *186*, 218–225. [CrossRef]

36. Reggi, S.; Giromini, C.; Dell'Anno, M.; Baldi, A.; Rebucci, R.; Rossi, L. In Vitro Digestion of Chestnut and Quebracho Tannin Extracts: Antimicrobial Effect, Antioxidant Capacity and Cytomodulatory Activity in Swine Intestinal IPEC-J2 Cells. *Animals* **2020**, *10*, 195. [CrossRef]

37. Sakagami, H.; Kuribayashi, N.; Iida, M.; Sakagami, T.; Takeda, M.; Fukuchi, K.; Gomi, K.; Ohata, H.; Momose, K.; Kawazoe, Y. Induction of DNA fragmentation by tannin- and lignin-related substances. *Anticancer Res.* **1995**, *15*, 2121–2128. [PubMed]

38. Bhat, R.; Hadi, S.M. DNA breakage by tannic acid and Cu(II) sequence specificity of the reaction and involvement of active oxygen species. *Mutat. Res.* **1994**, *313*, 39–48. [CrossRef]

39. Bhat, R.; Hadi, S.M. DNA breakage by tannic acid and Cu(ll) generation of active oxygen species and biological activity of the reaction. *Mutat. Res.* **1994**, *313*, 49–55. [CrossRef]

40. Mhlanga, P.; Perumal, P.O.; Somboro, A.M.; Amoako, D.G.; Khumalo, H.M.; Khan, R.B. Mechanistic Insights into Oxidative Stress and Apoptosis Mediated by Tannic Acid in Human Liver Hepatocellular Carcinoma Cells. *Int. J. Mol. Sci.* **2019**, *20*, 6145. [CrossRef]

41. Uddin, M.M.; Hasan, M.J.; Mahmud, Y.; Tuj-Zohra, F.; Ahmed, S. Evaluating Suitability of Glutaraldehyde Tanning in Conformity with Physical Properties of Conventional Chrome-Tanned Leather. *Text. Leather Rev.* **2020**, *3*, 135–145. [CrossRef]

42. St Clair, M.B.; Bermudez, E.; Gross, E.A.; Butterworth, B.E.; Recio, L. Evaluation of the genotoxic potential of glutaraldehyde. *Environ. Mol. Mutagen.* **1991**, *18*, 113–119. [CrossRef]

43. Zeiger, E.; Gollapudi, B.; Spencer, P. Genetic toxicity and carcinogenicity studies of glutaraldehyde-a review. *Mutat. Res. Rev. Mutat. Res.* **2005**, *589*, 136–151. [CrossRef]

44. Speit, G.; Neuss, S.; Schütz, P.; Fröhler-Keller, M.; Schmid, O. The genotoxic potential of glutaraldehyde in mammalian cells in vitro in comparison with formaldehyde. *Mutat. Res. Gen. Toxicol. Environ. Mutagen.* **2008**, *649*, 146–154. [CrossRef]

45. Inclusion of Substances of Very High Concern in the Candidate List for Eventual Inclusion in Annex XIV (Decision of the European Chemicals Agency). Available online: https://echa.europa.eu/documents/10162/f8bab22e-f605-bbed-66eb-13328354 36f9 (accessed on 7 September 2022).

46. Annex XV Report, Proposal for a Restriction, Substance Name(s): Chromium (VI) Compounds. Available online: https://echa.europa.eu/documents/10162/477b4727-e5fc-75da-ccd9-b8bcc2d2b7dd (accessed on 24 November 2021).

47. Błasiak, J.; Kowalik, J. A comparison of the in vitro genotoxicity of tri- and hexavalent chromium. *Mutat. Res. Gen. Toxicol. Environ. Mutagen.* **2000**, *469*, 135–145. [CrossRef]

48. Chiu, A.; Shi, X.L.; Lee, W.K.; Hill, R.; Wakeman, T.P.; Katz, A.; Xu, B.; Dalal, N.S.; Robertson, J.D.; Chen, C.; et al. Review of chromium (VI) apoptosis, cell-cycle-arrest, and carcinogenesis. *J. Environ. Sci. Health C Environ. Carcinog. Ecotoxicol. Rev.* **2010**, *28*, 188–230. [CrossRef]

49. Teklay, A. Physiological Effect of Chromium Exposure: A Review. *Int. J. Food Sci. Nutr. Diet.* **2016**, *S7*, 1–11. [CrossRef]

50. Klaude, M.; Eriksson, S.; Nygren, J.; Ahnstrom, G. The comet assay: Mechanisms and technical considerations. *Mutat. Res.* **1996**, *363*, 89–96. [CrossRef]

51. Uhl, M.; Helma, C.; Knasmuller, S. Single-cell gel electrophoresis assays with human-derived hepatoma (HepG2) cells. *Mutat. Res.* **1999**, *441*, 215–224. [CrossRef]

52. Uhl, M.; Helma, C.; Knasmuller, S. Evaluation of the single cell gel electrophoresis assay with human hepatoma (HepG2) cells. *Mutat. Res.* **2000**, *468*, 213–225. [CrossRef]

53. Lima, C.F.; Fernandes-Ferreira, M.; Pereira-Wilson, C. Phenolic compounds protect HepG2 cells from oxidative damage: Relevance of glutathione levels. *Life Sci.* **2006**, *79*, 2056–2068. [CrossRef] [PubMed]

54. Chi-Square, Cramer's V, and Lambda. Available online: http://vassarstats.net/newcs.html (accessed on 27 October 2021).

Article

Decontamination Efficiency of Thermal, Photothermal, Microwave, and Steam Treatments for Biocontaminated Household Textiles

Branko Neral *, Selestina Gorgieva 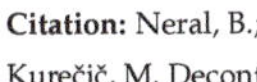 and Manja Kurečič

Faculty of Mechanical Engineering, University of Maribor, Smetanova Ulica 17, SI-2000 Maribor, Slovenia; selestina.gorgieva@um.si (S.G.); manja.kurecic@um.si (M.K.)
* Correspondence: branko.neral@um.si

Abstract: With the outbreak of the COVID-19 pandemic, textile laundering hygiene has proved to be a fundamental measure in preventing the spread of infections. The first part of our study evaluated the decontamination efficiency of various treatments (thermal, photothermal, and microwave) for bio contaminated textiles. The effects on textile decontamination of adding saturated steam into the drum of a household textile laundering machine were investigated and evaluated in the second part of our study. The results show that the thermal treatment, conducted in a convection heating chamber, provided a slight reduction in efficiency and did not ensure the complete inactivation of *Staphylococcus aureus* on cotton swatches. The photothermal treatment showed higher reduction efficiency on contaminated textile samples, while the microwave treatment (at 460 W for a period of 60 s) of bio contaminated cotton swatches containing higher moisture content provided satisfactory bacterial reduction efficiency (more than 7 log steps). Additionally, the treatment of textiles in the household washing machine with the injection of saturated steam into the washing drum and a mild agitation rhythm provided at least a 7 log step reduction in *S. aureus*. The photothermal treatment of bio contaminated cotton textiles showed promising reduction efficiency, while the microwave treatment and the treatment with saturated steam proved to be the most effective.

Keywords: household textiles; thermal treatment; decontamination; reduction efficiency

Citation: Neral, B.; Gorgieva, S.; Kurečič, M. Decontamination Efficiency of Thermal, Photothermal, Microwave, and Steam Treatments for Biocontaminated Household Textiles. *Molecules* **2022**, *27*, 3667. https://doi.org/10.3390/molecules27123667

Academic Editors: Ana Sutlović, Sanja Ercegović Ražić and Marija Gorjanc

Received: 17 May 2022
Accepted: 4 June 2022
Published: 7 June 2022

Publisher's Note: MDPI stays neutral with regard to jurisdictional claims in published maps and institutional affiliations.

1. Introduction

The process of caring for household textiles usually consists of the following phases: laundering, drying, ironing, and storage. The primary purpose of caring for household textiles is to ensure the highest level of textile cleanliness and hygiene without changes in the textiles' basic functions, such as comfort, functionality, and health protection.

Very often, it is forgotten that the conditions of the care process (e.g., temperature, time, water ratio/degree of humidity, concentration of detergent/bleaching agent/disinfectant, and mechanical action) must adapt to the physical–chemical characteristics of the textiles, such as the type of fibre and form (knitwear/fabric/nonwoven), the type of finish (bleached/dyed/printed/surface-functionalised), the level of hygroscopicity/hydrophobicity, the thermal conductivity, and the fastness type and level, and not the opposite. It seems to be entirely expected that the textile care process will not cause damage or shorten the life cycle of textiles and be environmentally sustainable. According to Bloomfield, textile laundering and drying are the most frequent household occupations, essential to ensuring hygiene, reducing the risk of infections, and, thus, protecting the health and safety of household members [1].

Statistics show that 82.5% of EU-28 households have a laundering machine, which is used 3.25 times/week, and 78.8% of households have a tumble dryer, which is used 2.05 times/week [2–4]. The annual energy consumption and water consumption of household laundering and drying machines in the EU-28 were estimated to be 35.3 TWh and 2.496 million m^3, respectively [5]. The European Environmental Policy, which rests upon the precautionary and prevention principles, projected annual electricity savings of 2.5 TWh, leading to GHG emission reductions of 0.8 Mt CO_2 eq/year and estimated water savings of 711 million m^3 in 2030 [6]. Clearly, EU institutions have public support for their ambitious plan [7].

In recent years, the number of studies investigating household laundering technology has been increasing [8–11]. Several studies have found that reducing the laundering time, the bath temperature, and water and electricity consumption may have positive financial impacts (a reduction in household textile laundering costs) and ecological impacts (a reduction in environmental contamination) at a comparable cleaning efficiency. However, most of these studies overlook the physical and chemical characteristics of different textiles, such as relations between factors of the Sinner circle, which determine the laundering quality.

In contrast, several studies have shown that laundering household textiles at temperatures below 40 °C, with a low bath ratio and a short laundering time, and without bleaching/disinfection agents may cause problems with the hygiene of textiles and provide low decontamination efficiency. Furthermore, the results indicate that the washing machine itself may serve as a vector for the contamination of textiles, the formation of biofilms, the development of unpleasant odours, and the growth of pathogenic microorganisms [12–14].

Overall, these studies demonstrate that the removal of bacteria, fungi, and viruses from textiles during the care process is influenced decisively by the mechanical and chemical mechanisms related to temperature, loading ratio, addition of bleaching agents, and duration [15–17]. It is also important to note that, while textile washing temperatures below 50 °C remove 95% of microorganisms [18,19], the rest are expected to be destroyed by tumble drying and ironing (particularly steam ironing).

However, far too little attention has been paid to textile drying and decontamination processes. Only a few research reports have been published that refer to industrial or household textile drying and decontamination.

According to Carr, the term 'drying' is generally used to describe the process of dewatering textiles, either mechanically (squeezing, vacuum extraction, centrifugation) or thermally (conduction, convection, infra-red, microwaves). Thermal textile drying is a complex process affected by the thermal and sorption properties of textiles, heat and mass transfer to and from the surface and inside of the textile fibre, the hydrodynamics of particle motion inside the fibre, and the various mechanisms by which moisture migrates through porous textile materials. The traditional thermal textile drying process is extremely energy-intensive and must be planned out and performed thoroughly [20–22].

Thermal textile drying in industrial laundries is performed with a drum dryer, a garment finisher, or a flatwork ironer depending on the form, the fibre type, and the surface functionalisation of the textile. Household textiles are usually dried in one of two ways: enhanced outdoor drying with wind and sunlight or indoor drying with a tumble dryer or drying cabinet [23–25].

For both industrial and household textile drying, the most critical parameter in the correctly performed process is the amount of moisture inside the textile material during, and especially at the end, of the drying process (moisture regain). Under-drying may result in the growth of microorganisms, whilst over-drying may result in a reduction in hygroscopic moisture that may cause a change in the fibre's morphology and, thus, a deterioration in product quality [21,26]. These factors can, consequently, shorten the lifespan of household textiles and, at the same time, increase emissions of textile wastes.

Upon analysing a few studies, Bloomfield found that textile drying after laundering can reduce the microbial load by between 0.4 and 4.5 log colony-forming units (CFUs), despite the significant variations in the data. The results suggest that drying at higher temperatures can produce further reductions in the numbers of microorganisms that survive the laundering process. Bloomfield also found that another factor likely to affect microorganisms' detachment is the extent to which contaminated fabrics are dried before laundering [27]. Together, the studies reviewed in the IFH 2011 Report show that the strength of microorganisms' adhesion to fabrics increases with drying [28].

The textile industry has investigated many uses for microwave energy, from heating, drying, dye-fixation, and the curing of resin-finished fabrics to disinvesting wool fabrics, but only from the heating point of view [29–33]. It can be concluded that significant progress has been made in the development and application of microwave technology in textile finishing processes, while its applicability to decontamination has been more or less overlooked.

Chen studied the applicability of a kitchen microwave oven for the decontamination of cotton inoculated with *Aspergillus niger,* a common mould (a type of fungus) that lives indoors and outdoors and can cause people with a weakened immune system to have health problems (e.g., allergic reactions, lung infections, and infections in other organs). The results of studies indicate that microwave irradiation has potential as a tool for textile decontamination; however, the limitations of conventional household microwave ovens (e.g., low efficiency, nonuniform heating, and a lack of a continuous source of moisture) need to be assessed, and at least partially rectified, in order to render microwave treatment a viable and practical tool for textile decontamination [34–36].

Gupta and Versteeg reviewed thirty-three studies from 1920 to 2016 to provide insight into footwear, sock, and textile sanitisation. They evaluated less-known techniques, such as UV light, silver-light (silver oxide (Ag_2O) activated by ultraviolet light), and ozone. The authors found that new approaches to shoe and sock sanitisation have yielded auspicious results, but further research and development are needed [37].

SARS-CoV-2 has spread to all continents in just a few weeks. It has been reaffirmed that basic hygiene measures, including textile hygiene, are the most effective non-pharmacological measures in limiting the spread of viruses [38–42]. Unfortunately, one of the features of the first wave of the COVID-19 pandemic was the lack of protective face masks and timely and accurate recommendations on the care and disinfection of textiles. Hospital and industrial laundries adapted their washing and disinfection processes quickly to the new conditions provided by the manufacturers of industrial detergents and disinfectants. However, significantly less information and fewer recommendations were available to households.

Bloomfield and Bockmühl concluded that textile laundering is the most frequent household occupation, essential to ensuring hygiene, reducing the risk of infection, and, thus, protecting the health and safety of household members [1,43]. This finding can also be linked to statistical data on SARS-CoV-2 infections in the Republic of Slovenia collected by the National Institute of Public Health from 1 November 2020 to 19 March 2022. Figure 1 shows the weekly number of infections depending on the most probable location of virus transmission, e.g., household, industry, hospital/healthcare institution, and nursing home.

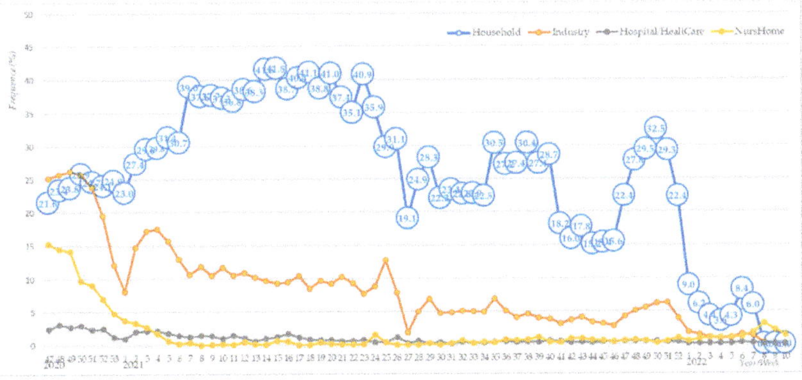

Figure 1. Weekly number of infections depending on the most probable location of virus transmission in the Republic of Slovenia for the period from 1 November 2020 to 19 March 2022 [44].

Figure 1 shows that the number of likely infections in workplaces and households at the end of 2020 was almost equal, slightly lower in nursing homes, and the lowest in hospitals. When the lockdown was introduced, the number of likely infections in workplaces fell sharply, in contrast to households, where it began to rise and then finally subsided in the first quarter of 2022.

In this study, we focused on less-investigated procedures for removing microbial contaminants from household textiles. In the first part of the study, we investigated and evaluated the effects of microorganism decontamination on textiles by the application of hot air, a combination of thermal and UV drying (photothermal treatment), microwaves, and saturated steam at atmospheric pressure. Analysis of the obtained results helped us to specify the conditions under which we can achieve the highest level of decontamination of textiles before or after household textile laundering and tumble drying.

2. Results and Discussion

The main objective of this research was to evaluate the effects of microorganism decontamination on cotton textile samples with different treatments that can be performed independently or as a part of household washing and drying processes. First, we evaluated the antibacterial efficiency of different drying processes, including classical thermal treatment, a combination of thermal and UV treatments (photothermal treatment), and, finally, microwave technology. The second part of the study was carried out by adding saturated steam to the washing drum.

The reason for the evaluation of the selected procedures for the removal of microorganisms from textiles was the outbreak of COVID-19, when Public Health Institutions, due to the lack of personal protective equipment, proposed that households use non-pharmacological measures to prevent the contamination and spread of SARS-CoV-2. Among other things, these institutions also proposed the use of home-made textile face coverings [45,46]. Unfortunately, the recommendations on the care and disinfection of face coverings were not supported by the results of the studies, and, at the same time, it was almost impossible to implement and evaluate the effects of disinfection in the household environment.

2.1. Thermal Treatment

Textile bioindicators, located on a holder in the middle of the drying chamber, were exposed to thermal treatment under different conditions in a laboratory dryer. The samples were dried for a period of between 30 and 90 min at 74 ± 0.71 °C, 82 ± 0.51 °C, and 91 ± 0.23 °C. The degree of *S. aureus* reduction was calculated based on an evaluation of the initial number of colonies and the number of colonies after the heat treatment. The results, in logarithmic step reductions in *Sa*, are shown in Figure 2.

From Figure 2, we can conclude that the degree of *S. aureus* reduction on the textile bioindicators increased with the temperature and heat treatment time, which was expected. The lowest degree of *S. aureus* reduction was achieved at 70 °C and 30 min (RED = 1.432), while 90 min of heat treatment at 90 °C yielded the highest degree of reduction (RED = 3.115). By increasing the time from 30 to 90 min at 70 °C, the reduction factor increased by 0.51, the same as at 80 °C, while at 90 °C, slight increases were observed to a value of 0.71.

We concluded that the low reduction rates were due to the testing conditions of the drying air, which was dormant (natural convection). Air circulation performs a vital transport function during laundry drying, as it supplies the heat and dissipates moisture from the textile material, which occurred to a minimal extent during the thermal treatment of the textile samples.

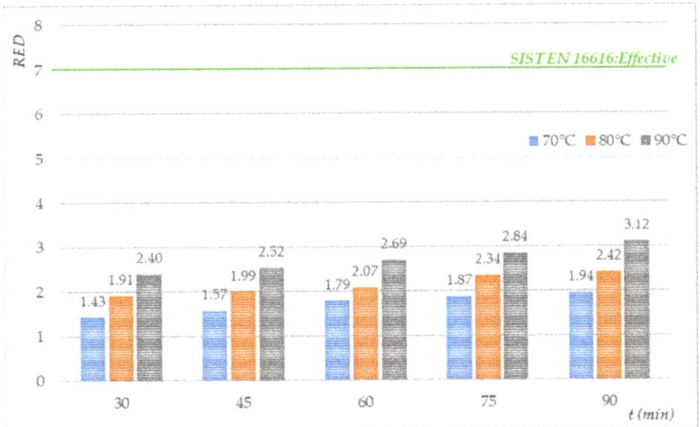

Figure 2. Log step reduction in *S. aureus* on textile bioindicators thermally treated in a heating oven (average $N_0 = 1.47 \times 10^7$ CFUs/mL).

Based on the results (not shown), we concluded that the fabric temperature was lower than the temperature of the drying air. Most of the heating energy was used to evaporate the surface water and the water between the threads and heat the fabric surface. The consequence of the low air circulation is reflected in the low degree of reduction in *S. aureus* on the textile swatches and in the temperature regulation in the heating chamber (temperature sensors, heaters, and electronic control system).

Therefore, the thermal treatment did not reach at least a 7 log step reduction in *S. aureus* on the cotton swatches. Despite this, we were interested in determining the conditions of drying in the heating chamber with natural air convection under which it would be possible to achieve the desired reduction in microorganisms. For this purpose, we used the GInaFit freeware software tool.

With the help of the GInaFIT V1.7 software, various mathematical models for the prediction of the inactivation of the microbial population were tested, from a simple log-linear model to the Weibull model. The use of the log-linear prediction model, which is a simple first-order model that represents exponential inactivation, gave a coefficient of determination (R^2) of 0.756. Meanwhile, in the Weibull prediction model, which represents the decline in microbial numbers as a cumulative distribution of heat lethality [47], the R^2 ranged from 0.9954 to 0.9987 for thermal treatment at 70 to 90 °C. The results are shown in Table 1, while the survival curves of *S. aureus* on textile swatches thermally treated in a heating oven are presented in Figure 3.

Table 1. The Weibull prediction model parameters for the heat resistance of *S. aureus* on textile swatches thermally treated in a heating oven.

T (°C)	N_0 (log CFUs/mL)	D_{value} (min)	t_{7D} (min)	RMSSE	R^2
70	7.25	33.91	237.37	0.034	0.9987
80	7.11	27.28	190.27	0.078	0.9954
90	7.11	21.58	151.06	0.083	0.996

N_0, initial value of *S. aureus* before thermal treatment; D_{value}, decimal reduction time; t_{7D}, time needed to reach a 7 log reduction in *S. aureus* at temperature T; RMSSE, Root Mean Sum of Square Errors; R^2, coefficient of determination.

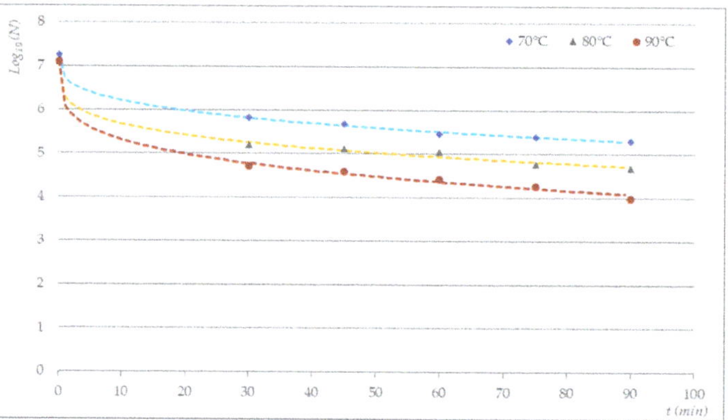

Figure 3. Measured and predicted surviving *S. aureus* microorganisms on textile bioindicators thermally treated in a heating oven based on the Weibull prediction model (GInaFIT v1.7).

As expected, the D_{value} decreased from 33.91 to 21.58 with respect to the temperature increase from 70 °C to 90 °C (Table 1). Based on the Weibull model, a 7 log step reduction would be achieved with heat treatment for 237.37 min at 70 °C, while heat treatment at 90 °C would yield the same reduction after 21.58 min.

Research on the applicability of a prediction model should be expanded to include other temperature-resistant microorganisms typically found on textiles in hospitals and nursing homes, such as temperature-resistant *Enterococcus faecium*. It is also necessary to consider the preparation and treatment conditions (pH, temperature, and moisture content) that could be used to predict the microorganism reduction efficiency of textile hygiene processes more reliably and eliminate long-term testing procedures [48–52].

2.2. Photothermal Treatment

A thermogravimetric moisture analyser was used to evaluate the antibacterial efficiency of the combined thermal and UV treatment of cotton swatches contaminated with *S. aureus*. The initial intense emission of UV waves and heat from the halogen lamp above the test samples was followed by the emission of shorter, less-severe waves of light and heat depending on changes in the mass of the dried sample. The treatment process was completed when the difference in mass between successive measurements was less than 0.01%. The drying profiles of the treated samples and the decontamination effects are presented in Table 2 and Figure 4, respectively.

Table 2. Influence of photothermal treatment on the survival of *S. aureus* on textile swatches.

T (°C)	t (min)	N (log CFUs/mL)	RED
-	0	8.301	-
50	53	6.982	1.319
55	37	6.663	1.638
60	33	6.465	1.836
65	26	6.029	2.272
70	20	5.664	2.637
75	14	4.655	2.910
80	10	4.193	3.372
85	7	3.756	3.809
90	4	3.556	4.008

T, treatment temperature; t, time of treatment set by the moisture analyser (diff <0.01% between two continuous measurements); N, average value of the surviving *S. aureus* after exposure to the treatment; RED, reduction factor.

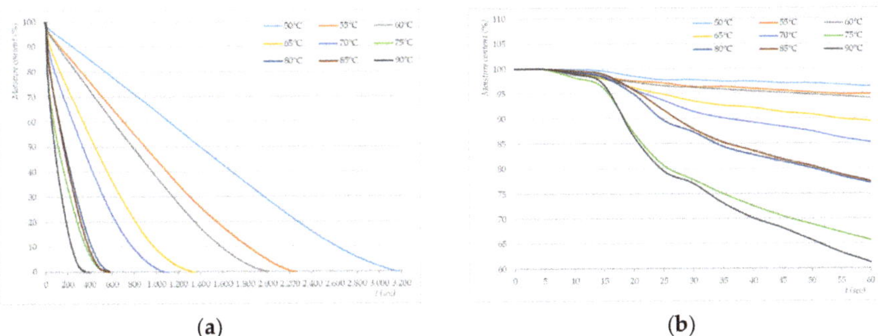

(a)   (b)

Figure 4. (**a**) Drying profiles of photothermally treated cotton swatches inoculated with *S. aureus* (average $N_0 = 1.18 \times 10^8$ CFU/mL); (**b**) magnification of the first 60 s of treatment.

The results (Table 2) show that the reduction in *S. aureus* increased with the temperature of the photothermal treatment; however, the treatment did not inactivate the entire population of *S. aureus* bacteria on the test samples. Thus, the difference between the highest reduction rate and the lowest reduction rate is 2.690 log steps. The highest reduction rate was provided by treatment at 90 °C for 4.008 log steps, while most of the *S. aureus* bacteria survived at 50 °C (6.982 log CFUs/mL). With the increase in the temperature of the photothermal treatment, the treatment time was shortened (Figure 4), and a lower number of surviving *S. aureus* bacteria on test samples was observed.

It is known that the process of drying cotton textiles with hot air has several stages. When heating the fibre by convection, a slight decrease in the moisture content (adhesive water) is characteristic of the initial phase of hot air drying (Figure 4b). This process is followed by the migration of water from inter-fibre spaces to the fibre surface (the hygroscopicity of the fibre surface for capillary water) and a rapid decline in the moisture content (Figure 4a). The formation of a plateau on the moisture content curve indicates the phase when only hygroscopic water remains in the fibre (the equilibrium moisture content, ECM). If hydroscopic water is removed due to an uneven moisture distribution, dehydration occurs, which damages the fibre (a reduction in dimensional stability, stiffness, smoothness, and breaking strength) [53–55].

Based on the results, it can be concluded that the microorganism reduction factor lags behind the drying profile in the photothermal treatment. This can be attributed to the phenomenon pointed out by Probstein and Donnarumma that the migration of moisture towards the fibre surface is affected by the surface tension of the water, including the concentration of surface-active agents and the surface potential [56,57]. In our case, however, it should be pointed out that the test samples were impregnated with a suspension of microorganisms before the photothermal treatment, which may have further contributed to the irregular migration of moisture and the inhomogeneous heating of the fibre. This finding is consistent with previous studies [52,58,59].

It is interesting to compare the drying profiles (80 °C) of a cotton sample wetted in water (total water hardness = 2.6 mmol/L) and a cotton sample contaminated with an *S. aureus* suspension (Figure 5). In the initial stages of the photothermal treatment (1–15 s) (Figure 5b), we observed no difference between the drying curves, both samples were heated evenly, and the moisture content did not change significantly. In the second phase of the drying process (Figure 5a), however, differences occurred when the moisture content began to decrease. The moisture content of the sample contaminated with *S. aureus* lagged behind the decrease in the moisture content of the samples soaked with water. This change occurs in the middle of the drying diagram (250 s). The moisture content of the sample contaminated with *S. aureus* decreased faster than that of the sample soaked with water. The decrease in the water content of samples soaked in water was almost linear and ended at 530 s. The reduction in the water content of the sample contaminated with *S. aureus* turned into a concave curve shape and ended at 575 s, which is 45 s later than in the sample wetted only with water. The diagram in Figure 5 shows another difference in the curves, indicating a change in the moisture content depending on the drying time. On the moisture content curve of the test samples soaked with *S. aureus*, several "stepped" sites can be observed. The occurrence of "stepped" parts of the curves was also observed in studies of the sorption properties of cotton samples by tensiometry [60]. These parts were due to the fact that it was difficult for water to migrate into the intermolecular areas of the fibre and inorganic/organic impurities that act as semipermeable membranes and affect the diffusion rate [61]. The observed phenomenon coincides with the findings of Hseih, Takashima, and Sanders, who reported that *Staphylococci spp.* adhered to cotton, PES, and blends of these materials to a greater extent compared with other microorganisms. They also stated that the fabric's water absorbency and saturation seem to increase the interactions between bacterial cells and the fibre [62–64].

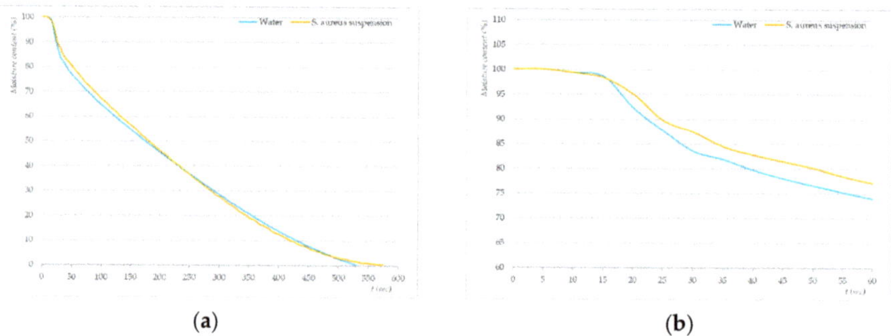

Figure 5. (**a**) Drying profiles of cotton swatches wetted with water (blue line) and swatches inoculated with a suspension of *S. aureus* ($N_0 = 3.67 \times 10^7$ CFU/mL, orange line) treated photothermally at 80 °C; (**b**) magnification of the first 60 s of the photothermal treatment.

2.3. Microwave Treatment

Microwave treatments were conducted in two series of tests (samples with a high moisture contest and samples with a low moisture content). In the first series of microwave treatments, test samples contaminated with an *S. aureus* suspension were used, followed by 2 h of drying in a safety chamber at room temperature. The moisture content of the test samples before drying was 85%, while after 2 h of drying in a safety chamber (22 °C, 30% RH) the moisture content was slightly lower (73%).

In the second series of tests, we were interested in how the lower moisture content of the bio contaminated textile samples would affect the antimicrobial efficiency of the microwave treatment, which is a common situation when preparing textile bioindicators for long-term use. Application of the suspension and 2 h of drying of the textile samples in a safety chamber were followed by additional drying in Petri dishes supported in a safety chamber at a temperature of 22 °C and an RH of 30%, followed by 24 h of storage in a refrigerator (8 °C, 40% RH). The average moisture content of the textile samples after 24 h of drying and 24 h of storage was 12%.

The samples from the first and second sets were treated in a microwave oven at 460 W for different treatment times (from 15 to 90 s). Because of the known non-uniformity of the microwave power distribution in the oven cavity (which causes localised hot or cold spots), all contaminated textile samples were cured at the same location in the oven cavity, which was determined as described in [65,66]. The results are shown in Figure 6.

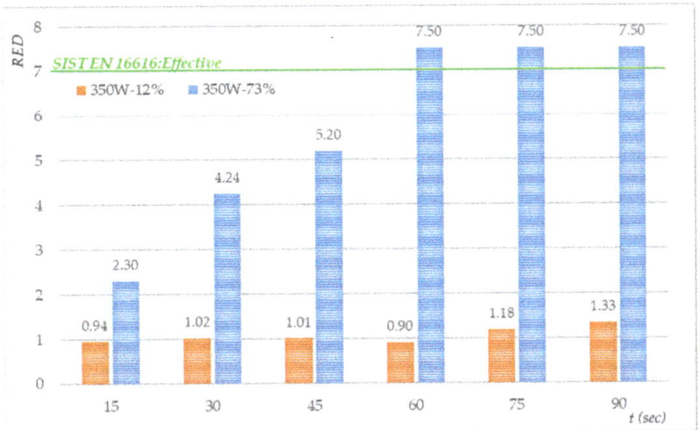

Figure 6. Antimicrobial efficiency of the microwave treatment of cotton swatches inoculated with *S. aureus* ($N_{0,350W,73\%MC} = 3.14 \times 10^7$ CFU/mL, $N_{0,350W,12\%MC} = 1.36 \times 10^7$ CFU/mL).

It can be concluded that the population of surviving *S. aureus* bacteria decreased by 1.36 log CFU/mL after 24 h of drying and 24 h of storage (a reduction from 7.497 to 6.134 log CFU/mL). Figure 6 shows that the reduction factor increased with the microwave treatment time, where a complete reduction in *S. aureus* (RED> 7.50 log steps) was achieved after 60 min. In contrast, the microwave treatment time at 460 W of test samples with 12% moisture content did not affect the increase in the RED factor and achieved a maximum reduction at 90 s of 1.33 log steps. The small effect on decontamination can be attributed to the low moisture content level and the small amount of heat generated during the microwave treatment.

The advantage of microwaves, which makes them attractive for use in textile hygiene processes, is their ability, under suitable conditions, to produce rapid and uniform heating throughout the material exposed to them. Microwave heating utilises the effect of a varying and high-frequency voltage on non-conductive (dielectric) material, which creates cyclic strains due to rapid changes in vibrational and rotational movements in the molecules. Under normal conditions, polar molecules of water, the most common polar molecules in textile fabrics, are oriented randomly. When wet textile fabric is exposed to alternating electromagnetic microwave energy, the polar molecules of water change their polarity rapidly and continuously and attempt to align themselves with the changing field (this is also known as the dipole heating mechanism). Friction thus arises from the molecules heating up and finally evaporating. The physics of this phenomenon are such that heating takes place only inside the wet parts of the fabric, and the heat is the most intense where the moisture content is the highest [32].

We also performed microwave tests on contaminated samples in the same way as Chen [34]; i.e., wetting with a suspension of *S. aureus* (130% moisture content), and then, without 2 h of drying, processing the samples with microwaves immediately. We found that microwave treatment of samples at 700 W for 30 s yields a complete reduction in *S. aureus* (RED> 7.50 log steps), which is desired. Less desirable, however, is the finding that there was damage to the cotton specimens (Figure 7). The visual observation of a change in the cotton swatches' colour from pastel, before the treatment, to yellow-brown, after the microwave treatment, indicates the consequences of thermodegradation of the cotton fibre.

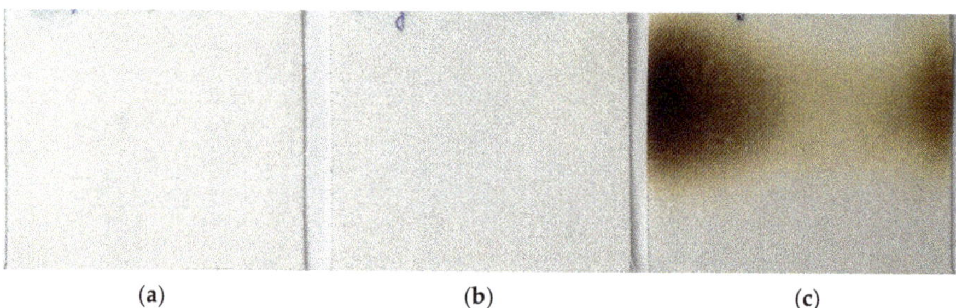

(a)   (b)   (c)

Figure 7. Damage to the cotton swatches after the microwave treatment; (**a**) 700 W, 10 s; (**b**) 700 W, 20 s; (**c**) 700 W, 30 s.

2.4. Steam Treatment

The treatment of contaminated textile samples was performed in a drum washing machine to which an external steam generator was connected. The injection of saturated steam (100–103 °C) into the gently agitated washing drum allowed for the homogeneous penetration of steam into all parts of the base load in the washing drum.

First, the temperature profile of the cotton base load was determined depending on the amount of steam added to three loads (30.77% (2.0 kg), 53.85% (3.5 kg), and 100% (6.5 kg)). Figure 8 shows that a 2.0 kg cotton base load reached 60 °C in 20 min by injecting saturated steam and cooled to 55 °C in 7 min. By adding steam at 30-min intervals, it was possible to maintain the temperature of 2.0 kg of ballast between 55 and 65 °C. A total of 3.5 kg and 6.6 kg of cotton ballast heated up significantly less and more slowly. Thus, 400 g of saturated steam was found to heat a 3.5 kg base load in 30 min at 45 °C, while, with 6.5 kg of cotton ballast, the initial temperature increased over the 30-min period by only a few degrees and did not exceed 20 °C. In preliminary research, we found that, for the sustainable and homogeneous heating of ballast, it is necessary to rotate it with a gentle agitation rhythm (3/12/52) and inject saturated steam continuously (400 g/30 min). These findings influenced our decision to carry out the subsequent phases with a 2.0 kg ballast load and processing times longer than 30 min. The results of the influence of the steam treatment in the laboratory washing machine on the reduction in *S. aureus* are shown in Figure 9.

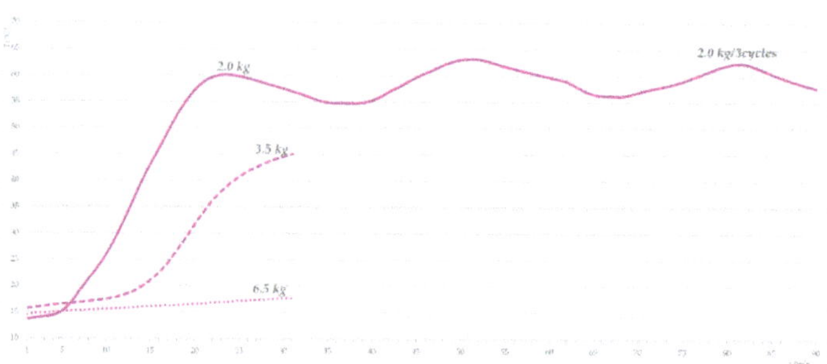

Figure 8. Temperature profile for different cotton base loads treated with saturated steam.

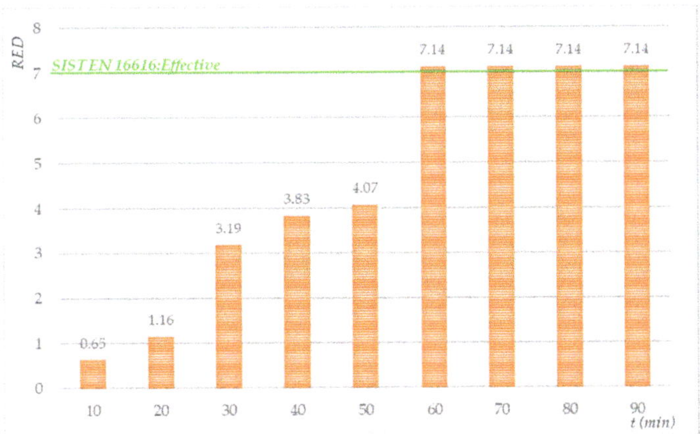

Figure 9. Antimicrobial efficiency of the injection of saturated steam into the drum of a laundering machine containing cotton swatches inoculated with *S. aureus* (average $N_0 = 2.56 \times 10^7$ CFU/mL).

Based on the results (Figure 9), we can conclude that the saturated steam treatment time affects the increase in the S. *aureus* reduction factor. A noticeable decrease in the population of surviving *S. aureus* bacteria on the textile swatches occurred after 30 min of treatment, while a complete reduction in *S. aureus* microorganisms was achieved after 60 min of treatment with saturated steam.

In the steam treatment, the moisture content in the base load increased from an initial 8.5% to 40.81%, 61.46%, and 85.74% after 40, 60, and 90 min, respectively (Figure 10). Saturated steam treatment could function as a stand-alone textile decontamination programme or a pre-programme in a washing programme. In the case of a combination with a washing programme, it would be possible to reduce the laundering machine's consumption of water significantly, as the base load will be wetted mainly by the pre-steam treatment.

Figure 10. Influence of the treatment with saturated steam (400 g/30 min) on the wetting of a cotton base load (2.0 kg).

3. Materials and Methods

We carried out procedures for the preparation of textile bioindicators, which, in the subsequent step, were treated thermally, photothermally, with microwaves, and with saturated steam for the purpose of evaluating the reduction efficiency of the treatments. A heating and microwave oven, a moisture analyser, a laboratory laundering machine, and a steam generator were used during the research. The bacterial reduction efficiency was measured and evaluated in accordance with ISO, EN, IEC, SIST, or VAH (The Association for Applied Hygiene (D)) requirements, specifications, or guidelines. All experiments were repeated three times.

3.1. Materials

Standard cotton fabric (SCF) was used as a microorganism carrier, and accompanying patterns were used to identify cross-contamination during the steaming process. In the research phase, when the effect of steam on decontamination was studied, a base load was used, consisting of IEC T11 cotton sheets, IEC T13 pillow cases, and IEC T12 towels. The characteristics of the SCF and the cotton base load items met the ISO, EN, and DIN standards and are shown in Table 3.

Table 3. Characteristics of the used fabric and base load items.

Parameter	Standard Cotton Fabric	Base Load		
Standard	DIN ISO 2267:2016	SIST EN 60456:2016, IEC PAS 62958:2015		
Form	Fabric	Bed sheet	Pillowcase	Towel
Fibre composition	100% cotton	100% cotton		
Density/Warp	27 threads/cm, 295 dtex	24 threads/cm, 330 dtex		24 threads/cm, 330 dtex
Density/Weft	27 threads/cm, 295 dtex	24 threads/cm, 330 dtex		12 threads/cm, 970 dtex
Mass	$170\ \text{g/m}^2$	$185\ \text{g/m}^2$		$220\ \text{g/m}^2$
Weave	Plain	Plain		Huckaback
Finish	Desizing, boiling, singeing, bleaching			
Colour	$C^*_{D65/10} = 0.65$, $h_{D65/10} = 98.62$, $WI_{CIE} = 71.69$	Not defined		

Before inoculation, the SCF was rinsed, dried, cut into swatches 1 m^2 in size, autoclaved, stored, and protected against recontamination, as is required in DIN ISO 2276:2016, IEC PAS 62,958:2015, and [67]. The ballast load was laundered in the washing machine and then dried in air, as defined in EN 60,456:2016.

The inlet water characteristics met the EN60,456:2016 standard (total water hardness (TWH), 2.5 ± 0.2 mmol/L; pH, 7.5 ± 0.2; T, 15 ± 2 °C).

All used textile materials and the IEC A* detergent were supplied by WFK Testgewebe GmbH (D).

3.2. Preparation of Textile Bioindicators

Textile bioindicators are used to determine the microorganism reduction efficiency of different treatments according to SIST EN 16,616:2015, IEC PAS 62,958:2015, and [67]. SCFs were used as carriers for the bacterial cultures of *Staphylococcus aureus DM 799*, which was obtained from the DSMZ German Collection of Microorganisms and Cell Cultures (D). The cotton swatches in the Petri dish were inoculated with 100 µL of artificial sweat that met the ISO 105-E04:2013 standard, as a substrate for simulating human excrement, and left to dry overnight. Sweat was chosen as it was previously found [68,69] that it can act as a substitute for defibrinated sheep blood. In the next step, 100 µL of a bacterial test suspension solution, previously prepared from stock culture as the first and second subcultures, was inoculated onto each cotton carrier. The carriers were then dried at room temperature in open Petri dishes for 2 h in a laminar flow cabinet and then used immediately or stored in the freezer. The initial concentration of bacteria on the cotton swatches was assessed by serial 10-fold dilutions and viable plate counting using Baird-Parker agar and found to lie between 10^7 and 10^8 per cotton piece.

3.3. Thermal, Photothermal, and Microwave Treatments

3.3.1. Drying and Heating Chamber

Textile bioindicators were treated thermally in an E-28 drying and heating chamber from Binder (D) at different temperatures and for different lengths of time. The main characteristics of the used chamber are a nominal power output of 0.8 kW, an interior chamber volume of 28 L, and dimensions of 400 mm × 280 mm × 250 mm. In one thermal treatment cycle, 5 pieces of the textile bioindicator, lying on a tray in the centre of the chamber, were processed. The air temperature during the thermal treatment was measured by a U12 Hobo single-channel temperature logger (Onset Computer Corp., Bourne, MA, USA) that can record up to 43,000 measurements. After the thermal treatment, the data were transferred to a PC and processed with MS Excel. The thermal treatment of the textile bioindicators was followed by an evaluation of the reduction in the population of the microorganisms in an ordinary way.

3.3.2. Moisture Analyser

A moisture analyser was used to determine the moisture content and photothermal inactivation of microorganisms. The basic principle of the analyser's operation is the rapid heating of a sample by heaters positioned above it and, consequently, the warming of the sample and the vaporisation of moisture. A precision electronic system determines the sample's weight continuously during the drying process. The process is completed when the system does not detect a difference in mass between consecutive measurements.

Usually, a moisture analyser device consists of a halogen lamp in which a mercury arc discharge is fired, a reflector system, a cooling system, and an electrical supply. An ordinary halogen lamp is composed of a quartz tube that contains mercury within an inert atmosphere. The lamp's body is made from quartz, ensuring that the UV energy has maximum transparency. The quartz body can resist an inner temperature of up to 1100 °C. Vacuum-tight tubes with sealed-in electrodes on both ends are used for the discharge of gas. The filling consists of an ignition gas (mostly argon) and liquid mercury spheres. A glow discharge is caused if the voltage above the cold electrodes is sufficient. When the electrodes are heated up, electrons are released from the cathode. This causes rapid growth in the number of electrons via shock ionisation of the filling gas. There is an arc discharge in the noble gas. Electrical energy is transferred to the light arc via the kinetic energy of the electrons. Impulses distribute the energy, the tube is heated up, and the mercury evaporates. The relative spectral density (RSD) of a halogen quartz glass heater is shown in Figure 11.

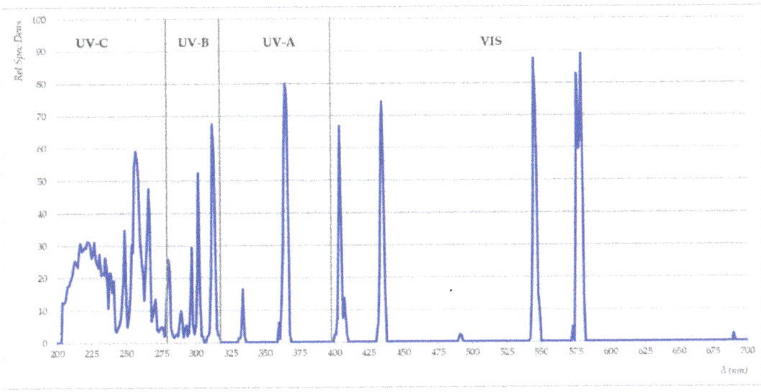

Figure 11. Relative spectral density (RSD) of a halogen quartz glass heater.

Just as important as the lamp itself is the system of reflectors that reflect the radiated light sideways or backwards onto the object. Using mirrors, the complete spectrum of short waves, visible light waves, and IR light waves are directed onto the object. The cooling of the halogen lamp with air is the responsibility of the control system (temperature sensors, microelectronic components, and a computer), which is the basis for the accurate control and fine-tuning of drying process conditions.

The KERN DBS 60-3 moisture analyser uses a thermogravimetric method. The sample is weighed before and after heating, and the moisture content in the material is determined by checking the difference. The moisture analyser's main characteristics are a power output of 400 W, a temperature range from 50 °C to 200 °C in steps of up to 1 °C, a 25 mm distance between the heater and the sample, and a sample plate made of fi90mm aluminium. The moisture content in the textile sample was calculated using Equation (1):

$$MC = (W_n - W_0)/W_0 \cdot 100 \tag{1}$$

where MC is the moisture content (%), W_0 is the initial weight of the sample before the drying step (g), and W_n is the weight of the sample (g) after the drying step.

3.3.3. Microwave Oven

Microwave curing was performed with an MO-17 ME microwave oven from Gorenje (SLO), which people use in their daily life to heat or warm food. The samples of textile bioindicators were microwave-cured at 460 W (cavity dimensions, 315 mm × 199 mm × 294 mm; capacity, 17 L) and a frequency of 2450 MHz over different lengths of time.

Details on the fundamentals of microwave heating and the construction and main parts of a household microwave oven can be found in [29].

3.4. Steam Treatment

The steam treatment procedures were conducted in a W365H laboratory washing machine from Electrolux (S) with a capacity of 6.5 kg, a drum volume of 65 L, and the possibility of programming the same mechanical action and duration using the programming controller Clarus Control and the PML Laundry Program Manager software, both from Electrolux (S), for customised laundering or drying procedures. The saturated steam needed for the treatment was produced by a 2331 Veit steam generator (D) (power output, 2.2 kW; steam output, 2.8 kg/h; boiler capacity, 5.9 L), which was connected to the washing machine. Figure 12 shows the scheme for the steam treatment system.

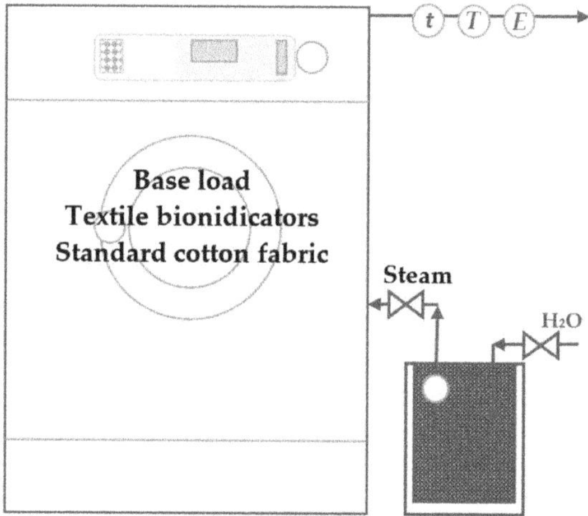

Figure 12. Scheme of the system for the saturated steam treatment.

The steaming procedure began with loading the washing machine with a base load (2.5 or 3.0 kg) of textile bioindicators and SCF inserted into different cotton laundering bags (5 pcs/steaming cycle), followed by rotation of the washing drum (gentle drum agitation,

3/12/52) and the injection of saturated steam through a nozzle (solenoid valve; T= 103 °C; p= 1.2 bar; spec. enthalpy= 2684 kJ/kg; steam connection= 9.5 mm). Treatments were performed under different conditions, lengths of time, and amounts of injected saturated steam.

Before each test run, the base load was hot-washed in a laundering machine (cotton; regular agitation; MW= 10 min/90 °C; RI1 and 2) with the addition of IEC A* detergent (base; SPT; TAED), dried, and stored prior to further use according to IEC PAS 62,958:2015.

Prior to each test run, the washing machine was conditioned. For that purpose, a self-cleaning programme was used, consisting of a main washing phase (45 min/85 °C) with a higher amount of water and four cold rinsing phases to cool down the laundering machine.

The electricity consumption was measured using the PowerSense Energomonitor (CZ) application. The temperature profile was measured by a U12 Hobo temperature logger placed among the base load items.

3.5. Evaluation of the Microbial Reduction

The concentration of microorganisms was evaluated on the cotton swatches during the incubation process and after the decontamination procedures. After each treatment, textile bioindicators were placed into a saline solution to release the cotton swatches. *CFU* was assessed by serial 10-fold dilutions and plating on selected agars. The average *CFU* after the incubation period (48 h/37 °C) was used to determine the reduction efficiency using Equation (2):

$$RED = \log (N_0/N) \tag{2}$$

where RED is the colony reduction factor efficiency, N_0 is the average CFU on the cotton swatches before the treatment procedure, and N is the CFU on the cotton swatches after the treatment procedure.

The usual limits for counting bacteria on agar plates are between 15 and 300; however, in SIST EN 16616:2015 and IEC 62958:2015 a deviation of 10% is accepted, so the counting limits were 14 and 330.

According to SIST EN 16616:2015, a chemical–thermal textile disinfection process is considered effective when a reduction rate of greater than 7 log steps is reached.

The microorganism inactivation rate caused by a treatment is expressed as the decimal reduction time D_{value}, which is used primarily in the food industry as a safety and quality evaluation parameter for thermal and nonthermal processes. It is known that the decimal reduction time is dependent on the temperature, the type of microorganism, and the composition of the medium containing the microorganisms [70]. The D_{value}, the time required to reduce the initial population of microorganisms by 90%, was calculated using Equation (3)

$$D_{value} = t/(\log N_0 - \log N_t) \tag{3}$$

where N_0 is the initial population of microorganisms (CFU/mL), N_t is the surviving population of microorganisms (CFU/mL), and t is the exposure time (min).

The GInaFiT V 1.7 freeware add-in for the MS Excel fitting tool, developed to test different types of microbial survival models on user-specific experimental data, was used in the research [71]. Due to the preliminary test results, the Weibull model was selected for the accurate prediction of microorganisms' inactivation.

4. Conclusions

The present study focused on less-investigated procedures for removing microbes from household textiles.

Gram-positive *S. aureus* is often selected as a test strain for the evaluation of disinfection properties of laundering procedures, detergents, and disinfectants in accordance with the recommendations of relevant standards. Therefore, it is no surprise that *S. aureus* was used instead of the COVID-19 virus for the development and evaluation of face mask disinfection procedures and agents at the beginning of the COVID-19 pandemic. In later studies, isolates of SARS-CoV-2 were used [17,72,73]. Based on our results, we can conclude that heat treatment in a convection heating chamber with an increased temperature and a longer treatment time affected the increase in the reduction factor but did not ensure a complete reduction in *S. aureus* on cotton swatches. The photothermal treatment of bio contaminated samples yielded higher reduction efficiency, but even this did not ensure the complete inactivation of the bacteria. In contrast, the microwave treatment of bio contaminated textile swatches, containing higher moisture contents, at 460 W for 60 s yielded satisfactory bacterial reduction efficiency. Additionally, the treatment of textiles in a household washing machine and the injection of saturated steam into the washing drum with a mild agitation rhythm yielded at least a 7 log step reduction in *S. aureus*.

When evaluating the effects of textile hygiene treatments, we must be aware that textiles are three-dimensional, flexible, and porous structures with specific sorption, dielectric, and heat-insulating properties. The periodic outbreak of epidemics proves that household textiles, caring for and the hygiene of household textiles, and the hygiene of washing/drying equipment play an essential role in preventing the spread of infections and protecting the health of household members. The change in the health situation must be followed by the development of washing and drying techniques for textile materials, but the principles of sustainable development must be considered too.

Author Contributions: Conceptualisation, methodology, investigation, resources, data curation, writing—original draft preparation, and supervision, B.N.; investigation, resources, data curation, and writing—review and editing, S.G.; investigation, resources, data curation, and writing—review and editing, M.K. All authors have read and agreed to the published version of the manuscript.

Funding: This research was funded by the Slovenian Research Agency (P2.14—Engineering sciences and technologies/Textile and Leather, applied project L2-3174) and by Gorenje d.o.o. Slovenia.

Institutional Review Board Statement: Not applicable.

Informed Consent Statement: Not applicable.

Data Availability Statement: Not applicable.

Acknowledgments: The authors would like to acknowledge Shelagh Margaret Hedges for the final proof-reading of the manuscript.

Conflicts of Interest: The authors declare no conflict of interest.

Sample Availability: Samples are not available from the authors.

References

1. Bloomfield, S.F. Importance of disinfection as a means of prevention in our changing world hygiene and the home. *GMS Krankenhhyg Interdiszip* **2007**, *2*, Doc25.
2. Pakula, C.; Stamminger, R. Electricity and water consumption for laundry washing by washing machine worldwide. *Energy Effic.* **2010**, *3*, 365–382. [CrossRef]
3. Michel, A.; Attali, S.; Bush, E. *Energy Efficiency of White Goods in Europe: Monitoring the Market with Sales Data—Final Report*; ADEME Press: Anger, France, 2016; p. 72.
4. Maya-Drysdale, L.; Iversen, N.H.; Gydesen, A.; Skov Hansen, P.M.; Gydesen, A. *Review Study on Household Tumble Driers—Final Report*; European Union: Brussels, Brussels, 2019; pp. 23–31.
5. Commission Regulation (EU) 2019/2023 of 1 October 2019 laying down ecodesign requirements for household washing machines and household washer-dryers pursuant to Directive 2009/125/EC of the European Parliament and of the Council, amending Commission Regulation (EC) No 1275/2008 and repealing Commission Regulation (EU) No 1015/2010 (OJ L 315, 5.12.2019, pp 28). Available online: https://eur-lex.europa.eu/legal-content/EN/TXT/?uri=CELEX%3A32019R2023 (accessed on 23 April 2021).
6. European Parliament. Environment Policy: General Principles and Basic Framework. Available online: https://www.europarl.europa.eu/about-parliament/en (accessed on 1 March 2022).
7. European Commission. State of the Energy Union 2021—Contributing to the European Green Deal and the Union's Recovery. Available online: https://ec.europa.eu/commission/presscorner/detail/en/ip_21_5554 (accessed on 23 March 2022).
8. Stamminger, R.; Barth, A.; Dörr, S. Washing Machines Wash Less Efficiently and Consume More Resources. *Hauswirtsch. Und Wiss.* **2005**, *53*, 124–131.

9. Gensch, C.O. Saving potentials by automatic dosage in washing machines. In Proceedings of the 44th IDC Conference Proceedings, Düsseldorf, Germany, 12–14 May 2009; WFK-Cleaning Technology Institute: Düsseldorf, Germany, 2009; pp. 68–73.

10. Barthel, C.; Götz, T. *What Users Can Save with Energy and Water Efficient Washing Machines-Final Report*; Wuppertal Institute for Climate, Environment and Energy: Wuppertal, Germany, 2013.

11. Abeliotis, K.; Amberg, C.; Candan, C.; Ferri, A.; Osset, M.; Owens, J.; Stamminger, R. Trends in laundry by 2030. *HPC Today* **2015**, *10*, 22–28.

12. Hammer, T.R.; Mucha, H.; Höfer, D. Infection Risk by Dermatophytes During Storage and After Domestic Laundry and Their Temperature-Dependent Inactivation. *Mycopathologia* **2010**, *171*, 43–49. [CrossRef] [PubMed]

13. Lucassen, R.; Blümke, H.; Born, L.; Fritz, A.; Geurtz, P.; Hofmann, N.; Hoffmann, L.; Steiner, R.; Merettig, N.A.; Bockmühl, D.I. The washing machine as a source of microbial contamination of domestic laundry—A case study. *Househ. Pers. Care Today* **2014**, *9*, 54–57.

14. Mucha, H.; Gerhardts, A.; Höfer, D. Pros and cons of hygiene claims in household laundry. In Proceedings of the 47th IDC Conference Proceedings, Düsseldorf, Germany, 19–21 May 2015; WFK-Cleaning Technology Institute: Düsseldorf, Germany, 2015; pp. 164–171.

15. Jaska, M.J.; Ames, L.; Fredell, D. Impact of detergent Systems on Bacterial Survival on Launderded Fabrics. *Appl. Environ. Microbiol.* **1980**, *39*, 743–748. [CrossRef]

16. Blaser, M.J.; Smith, P.F.; Cody, H.J.; Wang, W.L.; LaForce, F.M. Killing of Fabric-Associated Bacteria in Hospital Laundry by Low-Temperature Washing. *J. Infect. Dis.* **1984**, *149*, 33–41. [CrossRef]

17. Owen, L.; Laird, K. The role of textiles as fomites in the healthcare environment: A review of the infection control risk. *PeerJ* **2020**, *8*, e9790. [CrossRef]

18. Cuncliffe, V.; Gee, R.; Ainsworth, P. An investigation into some aspects of the efficiency of low-temperature laundering. *J. Consum. Stud. Home Econ.* **1988**, *12*, 95–106. [CrossRef]

19. Ainsworth, P.; Fletcher, J. A comparison of the disinfectatnt action of a powder and liquid detergent during low-temperature laundering. *J. Consum. Stud. Home Econ.* **1993**, *17*, 67–73. [CrossRef]

20. Carr, W.W.; Lee, H.S.; Ok, H. Drying of Textile Products. In *Handbook of Industrial Drying*, 3rd ed.; Mujumdar, A.S., Ed.; CRC Press: Boca Raton, FL, USA, 2006; pp. 781–791. [CrossRef]

21. Sousa, L.H.C.D.; Lima, O.C.M.; Pereira, N.C. Analysis of Drying Kinetics and Moisture Distribution in Convective Textile Fabric Drying. *Dry. Technol.* **2006**, *24*, 485–497. [CrossRef]

22. Gatarić, P.; Širok, B.; Hočevar, M.; Novak, L. Modeling of heat pump tumble dryer energy consumption and drying time. *Dry. Technol.* **2019**, *37*, 1396–1404. [CrossRef]

23. Berghel, J.; Brunzell, L.; Bengtsson, P. Performance analysis of a tumble dryer. In Proceedings of the 14th Int Drying Symposium (IDS 2004), São Paulo, Brazil, 22–25 August 2004; Volume B, pp. 821–827.

24. Stawreberg, L.; Berghel, J.; Renström, R. Energy losses by air leakage in condensing tumble dryers. *Appl. Therm. Eng.* **2012**, *37*, 373–379. [CrossRef]

25. Ambaritaetal, A. Performance of a clothes drying cabinet by utilizing waste heat from a split-type residential air conditioner. *Case Stud. Therm. Eng.* **2016**, *8*, 105–114. [CrossRef]

26. Ouyang, X.; Weiss, J.M.; de Boer, J.; Lamoree, M.H.; Leonards, P.E. Non-target analysis of household dust and laundry dryer lint using comprehensive two-dimensional liquid chromatography coupled with time-of-flight mass spectrometry. *Chemosphere* **2017**, *166*, 431–437. [CrossRef]

27. Bloomfield, S.F.; Exner, M.; Signorelli, C.; Scott, E.A. Effectiveness of Laundering Processes Used in Domestic (Home) Settings (2013) ISF on Home Hygiene. Available online: http://www.ifh-homehygiene.org/review/effectiveness-laundering-processes-used-domestic-home-settings-2013 (accessed on 2 August 2014).

28. Bloomfield, S.F.; Exner, M.; Signorelli, C.; Nath, K.J.; Scott, E.A. The Infection Risks Associated with Clothing and Household Linens in Home and Everyday Life Settings, and the Role of Laundry (2011) ISF on Home Hygiene. Available online: http://www.ifh-homehygiene.org/best-practice-review/infection-risks-associated-clothing-and-household-linens-home-and-everyday-life (accessed on 10 February 2019).

29. Haghi, A.K. Application of Microwave Techniques in Textile Chemistry—A Review. *Asian J. Chem.* **2005**, *2*, 639–654.

30. Katović, D.; Bischof Vukušić, S.; Flinčec Grgac, S. Aplication of Microwaves in Textile Finishing Processes. *Tekstil* **2005**, *54*, 313.

31. Pourová, M.; Vrba, J. Microwave Drying of Textile Materials and Optimization of a Resonant Applicator. *Acta Polytech.* **2006**, *5*, 3–7. [CrossRef]

32. Neral, B.; Šostar-Turk, S.; Schneider, R. Microwave fixation of ink-jet printed textiles. In Proceedings of the 4th International Textile, Clothing & Design Conference "Magic world of textiles", Dubrovnik, Croatia, 5–8 October 2008; Dragčević, Z., Ed.; Faculty of Textile Technology, University of Zagreb: Zagreb, Croatia, 2008; pp. 404–409.

33. Haggag, K.; Ragheb, A.; Nassar, S.H.; Hashem, M.; El-Sayed, H.; El-Thalouth, I.A. *Microwave Irradiation and its Application in Textile Industries*, 1st ed.; Science Publishing Group: New York, NY, USA, 2014; pp. 5–7.

34. Chen, H.L. Microwave Radiation Decontamination of Mildew Infected Cotton. *TRJ* **2001**, *71*, 247–254. [CrossRef]

35. Aspergillosis. Centers for Disease Control and Prevention. Available online: https://www.cdc.gov/fungal/diseases/aspergillosis/index.html (accessed on 29 April 2022).

36. Park, D.-K.; Bitton, G.; Melker, R. Microbial inactivation by microwave radiation in the home environment. *J. Environ. Health* **2006**, *69*, 17. [PubMed]
37. Gupta, A.K.; Versteeg, S.G. The Role of Shoe and Sock Sanitization in the Management of Superficial Fungal Infections of the Feet. *J. Am. Podiatr. Med. Assoc.* **2019**, *109*, 141–149. [CrossRef]
38. Leonas, K.K.; Jones, C.R. The Relationship of Fabric Properties and Bacterial Filtration Efficiency for Selected Surgical Face Masks. *JTATM* **2003**, *3*, 1–8.
39. MacIntyre, C.R.; Seale, H.; Dung, T.C.; Hien, N.T.; Nga, P.T.; Chughtai, A.A.; Rahman, B.; Dwyer, D.E.; Wang, Q. A cluster randomised trial of cloth masks compared with medical masks in healthcare workers. *BMJ Open* **2015**, *5*, e006577. [CrossRef]
40. Radonovich, L.J.; Simberkoff, M.S.; Bessesen, M.T.; Brown, A.C.; Cummings, D.A.; Gaydos, C.A.; Los, J.G.; Krosche, A.E.; Gibert, C.L.; Gorse, G.J.; et al. N95 Respirators vs. Medical Masks for Preventing Influenza Among Health Care Personnel, A Randomized Clinical Trial. *JAMA* **2019**, *322*, 824–833. [CrossRef] [PubMed]
41. Leung, N.H.; Chu, D.K.; Shiu, E.Y.; Chan, K.H.; McDevitt, J.J.; Hau, B.J.; Yen, H.L.; Li, Y.; Ip, D.K.; Peiris, J.S.; et al. Respiratory virus shedding in exhaled breath and efficacy of face masks. *Nat. Med.* **2020**, *26*, 676–680. [CrossRef] [PubMed]
42. Long, Y.; Hu, T.; Liu, L.; Chen, R.; Guo, Q.; Yang, L.; Cheng, Y.; Huang, J.; Du, L. Effectiveness of N95 respirators versus surgical masks against influenza: A systematic review and meta-analysis. *J. Evid. Based Med.* **2020**, *12*, 93–101. [CrossRef]
43. Bockmühl, D.P.; Schages, J.; Rehberg, L. Laundry and Textile Hygiene in Healthcare and Beyond. *Microb. Cell* **2019**, *6*, 299–306. [CrossRef] [PubMed]
44. Slovenian National Institute of Public Health (NIJZ). Available online: https://www.nijz.si/sl/dnevno-spremljanje-okuzb-s-sars-cov-2-covid-19 (accessed on 19 March 2022).
45. Community Face Coverings—Guide to Minimum Requirements, Methods of Testing and Use. Available online: https://www.cencenelec.eu/research/cwa/documents/cwa17553_2020.pdf (accessed on 28 September 2020).
46. Disinfection of Environments in Healthcare and Nonhealthcare Settings Potentially Contaminated with SARS-CoV-2. Available online: https://www.ecdc.europa.eu/en/publications-data/disinfection-environments-covid-19 (accessed on 15 April 2020).
47. Juneja, V.K.; Osoria, M.; Tiwari, U.; Xu, X.; Golden, C.E.; Mukhopadhyay, S.; Mishra, A. The effect of lauric arginate on the thermal inactivation of starved Listeria monocytogenes in sous-vide cooked ground beef. *Food Res. Int.* **2020**, *134*, 109280. [CrossRef]
48. Mazzola, P.G.; Penna, T.C.V.; da S Martins, A.M. Determination of decimal reduction time (D value) of chemical agents used in hospitals for disinfection purposes. *BMC Infect. Dis.* **2003**, *3*, 24. [CrossRef]
49. Bischel, H.N.; Schindelholz, S.; Schoger, M.; Decrey, L.; Buckley, C.A.; Udert, K.M.; Kohn, T. Bacteria Inactivation during the Drying of Struvite Fertilizers Produced from Stored Urine. *Environ. Sci. Technol.* **2016**, *50*, 13013–13023. [CrossRef] [PubMed]
50. Dementavicius, D.; Lukseviciute, V.; Gómez-López, V.M.; Luksiene, Z. Application of mathematical models for bacterial inactivation curves using Hypericin-based photosensitization. *J. Appl. Microbiol.* **2016**, *120*, 1492–1500. [CrossRef]
51. Channaiah, L.H.; Michael, M.; Acuff, J.C.; Lopez, K.; Phebus, R.K.; Thippareddi, H.; Milliken, G. Validation of a nut muffin baking process and thermal resistance characterization of a 7-serovar Salmonella inoculum in batter when introduced via flour or walnuts. *Int. J. Food Microbiol.* **2019**, *294*, 27–30. [CrossRef]
52. Lehotová, V.; Mihál-iková, K.; Medved'ová, A.; Valík, Ľ. Modelling the inactivation of Staphylococcus aureus at moderate heating temperatures. *Czech J. Food Sci.* **2021**, *39*, 42–48. [CrossRef]
53. Buisson, Y.L.; Rajasekaran, K.; French, A.D.; Conrad, D.C.; Roy, P.S. Qualitative and Quantitative Evaluation of Cotton Fabric Damage by Tumble Drying. *TRJ* **2000**, *70*, 739–743. [CrossRef]
54. Soljačić, I.; Pušić, T. *Njega Tekstila -Čišćenje u Vodenim Medijima*; Tekstilno-tehnološki fakultet Sveučilišta u Zagrebu: Zagreb, Croatia, 2005; pp. 109–113.
55. Wei, Y.; Su, Z.; Wang, Z.; Yuan, H.; Li, C. Development of an efficient and environmental-friendly drying model for domestic dryer. *J. Text. Inst.* **2020**, *111*, 214–225. [CrossRef]
56. Probstein, R.F. *Physicochemical Hydrodynamics: An Introduction*; John Wiley & Sons: New York, NY, USA, 2005; pp. 305–359.
57. Donnarumma, D.; Tomaiuolo, G.; Caserta, S.; Guido, S. Liquid flow in porous media. In Proceedings of the COST MP1106: Wetting of solids: Material and Kinetic Aspects, Cargese, France, 13–14 November 2013.
58. Shebuski, J.R.; Vilhelmsson, O.; Miller, K.J. Effects of growth at low water activity on thermal tolerance of Staphylococcus aureus. *J. Food Prot.* **2000**, *63*, 1277–1281. [CrossRef]
59. Zhang, L.; Kou, X.; Zhang, S.; Cheng, T.; Wang, S. Effect of water activity and heating rate on Staphylococcus aureus heat resistance in walnut shells. *Int. J. Food Microbiol.* **2018**, *266*, 282–288. [CrossRef]
60. Neral, B.; Mihelič, A.; Lesjak, P.; Habjanič Doler, I. *ColdCare-Optimization of Laundering Parameters and Cleaning Efficiency Improvement for HWM ASKO W4086C.W Gorenje d.o.o. Research Report*; University of Maribor, Faculty of Mechanical Engineering, Institute of Engineering Materials and Design: Maribor, Slovenia, 2018.
61. Pucci, M.F. Dynamic Wetting of Molten Polymers on Cellulosic Substrates: Model Prediction for Total and Partial Wetting. *Front. Mater.* **2020**, *7*, 143. [CrossRef]
62. Hsieh, Y.L.; Timm, D.A.; Merry, J. Bacterial Adherence on Fabrics by a Radioisotope Labeling Method. *TRJ* **1987**, *57*, 20–28. [CrossRef]
63. Takashima, M.; Shirai, F.; Sageshima, M.; Ikeda, N.; Okamoto, Y.; Dohi, Y. Distinctive bacteriabinding property of cloth materials. *Am. J. Infect. Control* **2004**, *32*, 27–30. [CrossRef] [PubMed]

64. Sanders, D.; Grunden, A.; Dunn, R.R. A review of clothing microbiology: The history of clothing and the role of microbes in textiles. *Biol. Lett.* **2021**, *17*, 20200700. [CrossRef]

65. Gaukel, V.; Siebert, T.; Erle, U. Microwave-assisted drying. In *The Microwave Processing of Foods*; Woodhead Publishing: Amsterdam, The Netherlands, 2017; pp. 152–178.

66. Chandan, K.; Azharul, K. Microwave-convective drying of food materials: A critical review. *Crit. Rev. Food Sci. Nutr.* **2019**, *59*, 379–394.

67. Requirements and Methods for VAH Certification of Chemical Disinfection Procedures. In *The Association for Applied Hygiene-VAH, VAH Disinfectants Commission*, 1st ed.; mhp-Verlag GmbH: Wiesbaden, Germany, 2016.

68. Fijan, S.; Koren, S.; Cencič, A.; Šostar-Turk, S. Antimicrobial disinfection effect of a laundering procedure for hospital textiles against various indicator bacteria and fungi using different substrates for simulating human excrements. *Diagn. Microbiol. Infect. Dis.* **2007**, *57*, 251–257. [CrossRef]

69. Fijan, S.; Škerget, M.; Knez, Ž.; Šostar-Turk, S.; Neral, B. Determining the disinfection of textiles in compressed carbon dioxide using various indicator microbes. *J. Appl. Microbiol.* **2012**, *112*, 475–484. [CrossRef] [PubMed]

70. McCormik, P.J.; Finocchario, C.J.; Kaiser, J.J. Sterility and bioindicators. In *Book Microbial Contamination Control in Parenteral Manufacturing*; Wiliams, K., Ed.; Taylor & Francis: New York, NY, USA, 2004; pp. 100–125.

71. Geeraerd, A.H.; Valdramidis, V.P.; Van Impe, J.F. GInaFiT, a freeware tool to assess non-log-linear microbial survivor curves. *Int. J. Food Microbiol.* **2005**, *102*, 95–105. [CrossRef]

72. De Man, P.; van Straten, B.; van den Dobbelsteen, J.; Van Der Eijk, A.; Horeman, T.; Koeleman, H. Sterilization of disposable face masks by means of standardized dry and steam sterilization processes; an alternative in the fight against mask shortages due to COVID-19. *J. Hosp. Infect.* **2020**, *105*, 356–357. [CrossRef]

73. Saini, V.; Sikri, K.; Batra, S.D.; Kalra, P.; Gautam, K. Development of a highly effective low-cost vaporized hydrogen peroxide-based method for disinfection of personal protective equipment for their selective reuse during pandemics. *Gut Pathog.* **2020**, *12*, 29. [CrossRef]

Article

Influence of Hydrogen Peroxide on Disinfection and Soil Removal during Low-Temperature Household Laundry

Petra Forte Tavčer [1,*], Katja Brenčič [1], Rok Fink [2] and Brigita Tomšič [1]

[1] Faculty of Natural Sciences and Engineering, University of Ljubljana, Aškerčeva 12, 1000 Ljubljana, Slovenia; katja.brencic@ntf.uni-lj.si (K.B.); brigita.tomsic@ntf.uni-lj.si (B.T.)
[2] Faculty of Health Sciences, University of Ljubljana, Zdravstvena pot 5, 1000 Ljubljana, Slovenia; rok.fink@zf.uni-lj.si
* Correspondence: petra.forte@ntf.uni-lj.si

Abstract: In the Water, Energy and Waste Directive, the European Commission provides for the use of household washing programmes with lower temperatures (30–40 °C) and lower water consumption. However, low washing temperatures and the absence of oxidising agents in the liquid detergents, and their reduced content in powder detergents, allow biofilm formation in washing machines and the development of an unpleasant odour, while the washed laundry can become a carrier of pathogenic bacteria, posing a risk to human health. The aim of the study was to determine whether the addition of hydrogen peroxide (HP) to liquid detergents in low-temperature household washing allows disinfection of the laundry without affecting the properties of the washed textiles even after several consecutive washes. Fabrics of different colours and of different raw material compositions were repeatedly washed in a household washing machine using a liquid detergent with the addition of 3% stabilised HP solution in the main wash, prewash or rinse. The results of the antimicrobial activity, soil removal activity, colour change and tensile strength confirmed the excellent disinfection activity of the 3% HP, but only if added in the main wash. Its presence did not discolour nor affect the tensile strength of the laundry, thus maintaining its overall appearance.

Keywords: textile care; low-temperature washing; laundry hygiene; hydrogen peroxide; disinfection activity; soil removal; colour difference; tensile properties

Citation: Tavčer, P.F.; Brenčič, K.; Fink, R.; Tomšič, B. Influence of Hydrogen Peroxide on Disinfection and Soil Removal during Low-Temperature Household Laundry. *Molecules* **2022**, *27*, 195. https://doi.org/10.3390/molecules27010195

Academic Editors: Ana Sutlović, Marija Gorjanc and Sanja Ercegović Ražić

Received: 29 November 2021
Accepted: 27 December 2021
Published: 29 December 2021

Publisher's Note: MDPI stays neutral with regard to jurisdictional claims in published maps and institutional affiliations.

1. Introduction

Among the measures to reduce greenhouse gas emissions from households and to reduce energy consumption, the European Commission has mandated in the Water, Energy and Waste Directive that household washing programs with lower temperatures (30–40 °C) and lower water consumption must be used [1]. Given the huge amounts of laundry we wash in households today (clothing consumption is constantly increasing due to the rise of fast fashion), such an approach makes sense and can actually bring measurable savings in energy and water consumption [2,3]. LCA studies of clothing, detergents and washing machines show that the use phase is generally the most energy consuming phase in the life cycle of clothing, ahead of the production and transport phases [4]. Consumer washing habits vary [5,6], but in the last 20 years the preferred washing temperatures worldwide have been between 30 °C and 60 °C [4–8]. However, these washing methods cannot guarantee the sufficient cleanliness of the laundry and especially the sufficient disinfection of the laundry [9]. It is worth noting that any pathogen associated with human disease is likely to be found on textile clothing [10]. Since these are environmental pathogens, contamination of textile surfaces could have occurred after or during the laundering process. Accordingly, inadequate domestic laundry hygiene can be of great concern to certain at-risk groups (e.g., immunocompromised individuals, the elderly, young children, etc.), as contamination or recontamination from hygienically inadequately cleaned textiles or even cross-contamination between household members could cause certain health problems [11–14]. It has been found that hygienic textiles can only be achieved at a washing temperature above 60 °C [15]. Accordingly, it is realistic to assume that potentially pathogenic microorganisms are not eliminated when infected laundry is washed at temperatures of 30–40 °C, especially since these temperatures are ideal for their development and multiplication [7,11,16].

The problem of textile washing hygiene is partly solved by oxidizing agents added to detergents in the form of perborates or percarbonates (bleach activators) [9,11,17]. Under suitable conditions, active oxygen is released into the wash bath where it oxidises coloured stains while acting against a variety of microorganisms (bacteria, bacterial spores, fungi, yeasts and viruses) [18–20]. Bleach activators are found only in solid detergent powders, but their share has been decreasing since 2010 due to EU environmental regulations. Liquid detergents, which are popular among consumers, do not contain bleach activators at all because they are unstable in liquid formulations [21]. Therefore, the result of using liquid detergents with a low-temperature laundry care program manifests itself in hygienically unsafe textiles and internal parts of the washing machine, which offers the possibility of spreading potentially pathogenic microorganisms through textile products [15,22,23]. Achieving adequate hygiene of laundry and internal parts of the washing machine during low-temperature washing is a particular problem with regard to current and future epidemics [24].

In order to achieve a balance between energy saving, adequate hygiene and effective soil removal, it is crucial to optimise the process of household laundry by introducing cold disinfection. Accordingly, the use of ecologically acceptable agents that do not pollute wastewater, are safe for humans and do not pose a risk for the development of long-term resistance of microorganisms due to increased use is urgently needed.

Among antimicrobial agents, hydrogen peroxide (HP) and peroxyacetic acid (PAA) are among the most environmentally friendly disinfectants [25,26]. They are characterised by high efficiency, broad spectrum antimicrobial activity, low toxicity and ease of use. HP is an antiseptic and antibacterial agent against various gram-negative and positive bacteria. Since HP efficacy is concentration-dependent, the presence of catalase and peroxidases in the cell can substantially decrease the antibacterial effect, and therefore concentrations below 3% are less effective [27]. Studies in the field of ensuring effective disinfection of laundry during the washing process with the addition of HP are limited in number and mostly oriented in the field of professional textile care [28]. Accordingly, the disinfection activity of the HP/PAA mixture of hospital laundry when washed at 40 °C was well studied [29]. Washing at 40 °C was found to reduce energy consumption while showing adequate disinfection activity against selected microorganisms, *Enterococcus faecium*, *Staphylococcus aureus*, *Enterobacter aerogenes* and *Candida albicans*, but at a higher concentration of HP/PAA. Compared to washing at 60 °C, washing at 40 °C showed the least mechanical damage of the standard cotton control. The wastewater showed some degree of soiling and was biodegradable, thus suitable for biological treatment. The antibacterial activity of HP during cold washing at 30 °C was also investigated using *Enterococcus faecium* and *Enterobacter aerogenes* as model thermoresistant Gram-positive and Gram-negative bacteria. The result showed excellent growth reduction of both tested bacteria, but the latter was time-dependent. Accordingly, the highest antibacterial activity was obtained during the main wash with a duration of 43 min [28]. In addition to bactericidal and fungicidal activity, the efficacy of HP against *Clostridiodes dificile* spores in the industrial tunnel washing of hospital laundry was also demonstrated, showing their sporicidal activity [30].

To the best of our knowledge, there is no study on the overall effect of HP addition in low-temperature household laundry (30–40 °C), which would provide systematic data on the laundry hygiene and stain removal efficiency, and its influence on colour changes and mechanical properties of textiles. Therefore, despite the extreme importance of providing hygienically adequate textile surfaces during household washing, which is one of the effective means of controlling the spread of pathogenic microorganisms and the occurrence of diseases, the introduction of HP in the household washing process remains an important, complex and largely unexplored research task. Accordingly, the aim of this study was to determine whether the addition of HP in the form of a 3% solution to a liquid detergent provides efficient disinfection and effective textile care during the household washing at 40 °C, and whether the use of 3% HP affects the properties of textiles. Moreover, the influence of the wash cycle on the overall effect of the 3% HP addition was also monitored. For this purpose, 3% HP was added during the main wash cycle, the prewash or during the rinse cycle. For each wash cycle, the disinfection activity was determined. The same washing procedures were used to test the removal of four different EMPA standard soils to give an insight on the stain removal efficiency. Influence on colour differences and tensile strength of the laundry were studied after consecutive washings of the randomly selected textile samples of different composition and colour with and without the addition of 3% HP in each wash cycle.

2. Results and Discussion

2.1. Disinfection Activity of 3% HP

Prior to testing the influence of the HP addition in the washing process on the appearance and performance of textile goods, its disinfection activity was investigated. The focus was on determining the most appropriate wash cycle for the HP addition that would reflect the highest disinfection efficiency. Accordingly, 3% HP was added during the wash, i.e., prewash, main wash and rinse. The results are presented in Figures 1 and 2. As can be seen in Figure 1, the addition of 3% HP caused some disinfection activity when added at the beginning (prewash) or at the end (rinse) of the wash, resulting in a 2.7–2.9 log reduction in the growth of the tested bacteria. On the other hand, the addition of 3% HP in the main wash cycle showed excellent inactivation of the tested bacteria, corresponding to more than 99.9999% reduction in bacterial growth. Accordingly, the highest disinfection activity was obtained against *E. coli* with a growth reduction of 7.62 log, followed by *P. aeruginosa* with a reduction of 6.61 log and *S. aureus* with a reduction of 6.58 log. The extreme increase in disinfection activity of 3% HP in the main wash compared to the prewash and rinse can be explained by the longer contact time of 3% HP with the tested bacteria, as well as higher temperature. Namely, in comparison to the prewash and rinse, the duration of the main wash was about 2.5 and 5 times longer, with 12–14 °C higher temperature of the washing bath. Accordingly, the prolonged duration of the main wash allowed a longer contact time of the tested bacterial cells with the extremely biocidal hydroxyl radicals (OH$^\bullet$) formed during the activation of HP in the presence of trace amounts of transition metal ions (known as the Fenton reaction) present in tap water, while higher temperature conditioned the faster activation reaction of HP [31,32]. It is assumed that generated OH$^\bullet$ reacts with bacterial cell lipids, proteins and nucleic acids to result in the break of RNA and DNA and the destruction of sulfhydryl bonds in proteins and membranes [33]. Since the rinse cycle of the wash was the shortest, one would expect the lowest disinfection activity in this case. However, the results show comparable disinfection activity of the HP between the prewash and rinse cycles. Since the cotton carriers of the tested bacteria were added at the beginning of the washing process, the disinfection activity of the HP in the rinse cycle was also partially achieved by the antimicrobial activity of the quaternary ammonium compounds contained in the liquid detergents [34]. In this case, the entire washing process without adding the 3% HP resulted in 1.81, 0.22 and 2.59 log reduction of *E. coli*, *P. aeruginosa* and *S. aureus*, respectively (data not shown). Similar to our study, Hooker et al. [35] demonstrated that adding the HP to the main wash can substantially reduce *E. coli*, *S. aureus* and *P. aeruginosa* (log 6.9) when washing hospital mattresses.

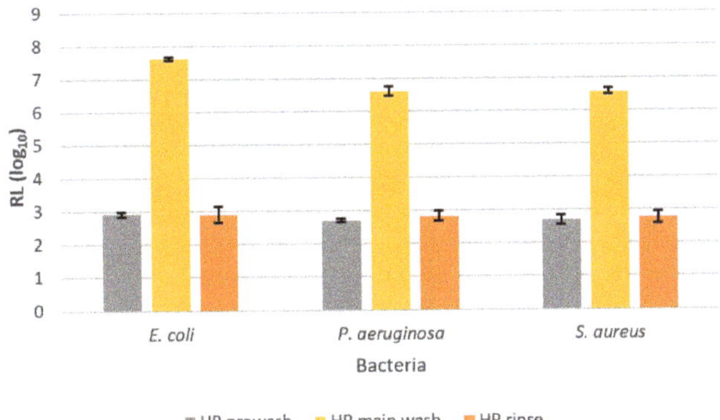

Figure 1. Disinfection activity, RL (log CFU cm^{-2}), of 3% hydrogen peroxide added in the prewash, main wash and rinse cycles against tested bacteria *E. coli*, *P. aeruginosa* and *S. aureus*.

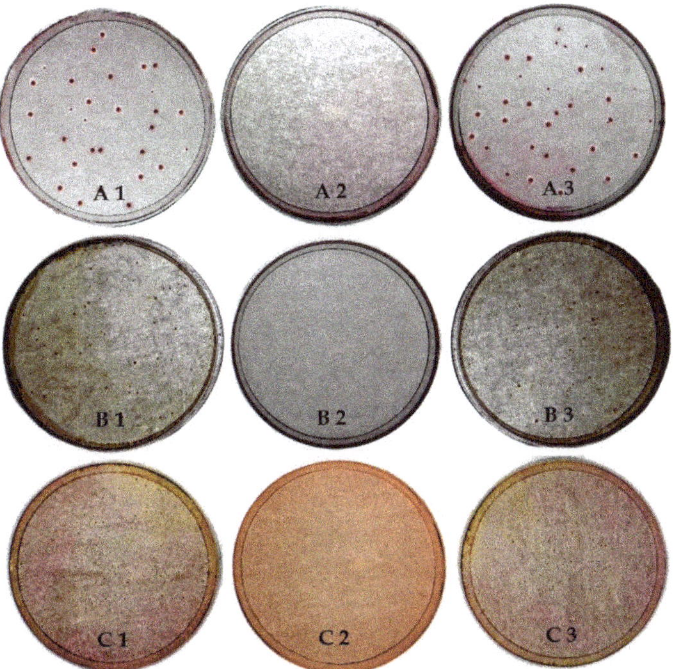

Figure 2. Agar plates of *E. coli* (**A**), *P. aeruginosa* (**B**) and *S. aureus* (**C**) growth after studied wash cycles in the presence of 3% HP as disinfectant agent: (**1**) prewash, (**2**) main wash and (**3**) rinse.

2.2. Standard Soil Removal

Figure 3 shows the difference in lightness of standard soiled fabrics, ΔLs*, with the addition of 3% HP at different washing cycles. The results show that all standard soils, with the exception of Empa 116 soil (blood/milk/ink), are washed better in the presence of HP. The addition of 3% HP to the main wash is most effective, followed by the prewash and finally the rinse. These results were to be expected, since the oxidation reaction of HP is highly temperature and time-dependent. As mentioned above, both the temperature and time of these wash cycles were lower and shorter than for the main wash. However, unlike the disinfection activity results, where comparable antimicrobial activity was found between the prewash and rinse, this was not the case for soil removal activity.

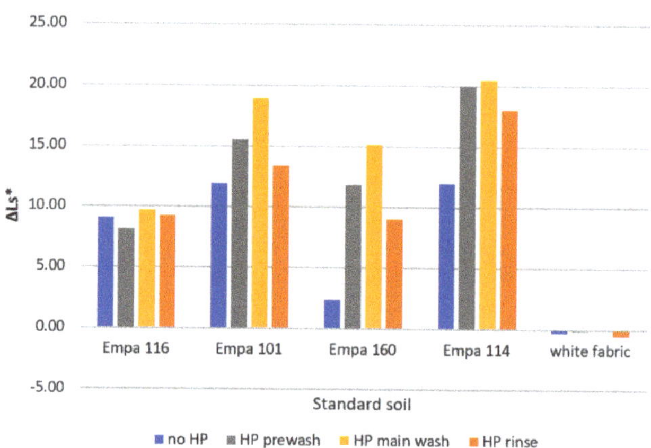

Figure 3. Lightness difference, ΔLs*, of standard soiled fabrics washed at 40 °C with liquid detergent Ox without (no HP) and with the addition of HP in studied wash cycles.

To gain better insight into the soil removal efficiency of 3% HP in each wash, the ratio between the difference in ΔLs* value between the sample of standard soil washed without 3% HP and with it was determined and is shown in Figure 4. Undoubtedly, the addition of 3% HP in the main wash cycle contributed the most to the removal of the studied soils. It was most active against Empa 160 (chocolate cream), followed by Empa 114 (red wine), Empa 101 (carbon black/olive oil) and finally Empa 116 (blood/milk/ink). Thus, the presence of 3% HP in the main wash greatly improved the removal of organic colour pigments into colourless due to its bleaching effect. In the case of standard soil Empa 116, it should be noted that due to the presence of haemoglobin in blood, a reaction occurs between iron and hydrogen peroxide, which leads to oxidation [31,32,36]. As the latter is conducive to disinfection activity, it impairs the formation of bleach ions, which is reflected in lower lightness and poorer soil removal. Accordingly, the addition of 3% HP in the prewash did not contribute to the improved removal of Empa 116 soil, but even slightly affected the washing performance in this wash cycle, as the obtained contribution value was negative. Nevertheless, better removal of this type of soils is obtained at higher pH and temperature [36]. The photos of the studied standard soils without and with the addition of 3% HP in the studied washing cycles are shown in Figure 5.

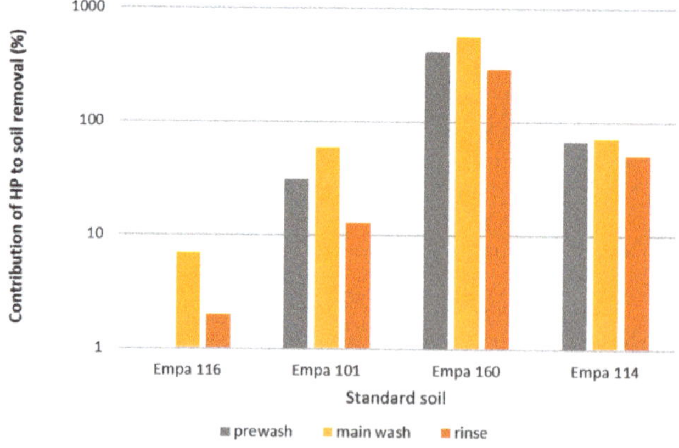

Figure 4. Contribution of HP [a] to soil removal after washing in the studied wash cycles. [a] Contribution of HP was determined as a ration between ΔLs* of the standard soil sample washed with 3% HP and ΔLs* of the same standard soil sample washed without 3% HP.

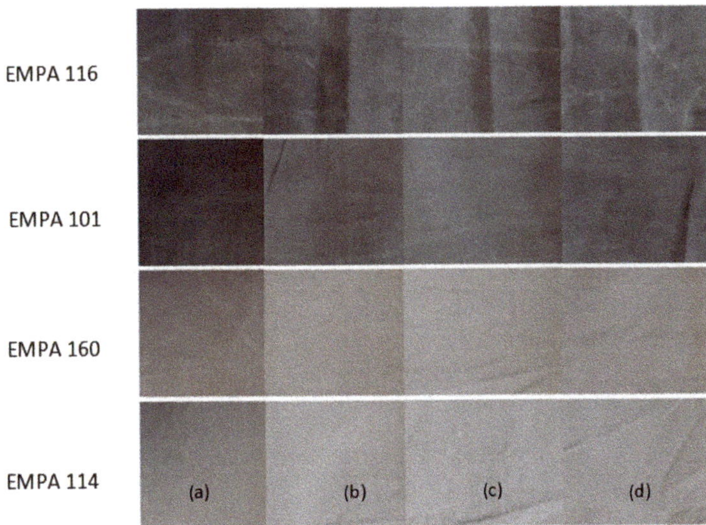

Figure 5. Photos of standard soils after washing without and with the addition of 3% HP in studied wash cycles: (**a**) No HP; (**b**) HP prewash; (**c**) HP main wash; (**d**) HP rinse.

2.3. Colour Change of Textile Samples

Figure 6 shows the colour changes of textile samples washed five consecutive times with 3% HP in the studied wash cycles. The results show that 3% HP does not affect the colour of textiles, as the colour difference ΔE_c^* is less than 1.5, which is estimated to be an imperceptible colour difference. The greatest colour changes were expected when hydrogen peroxide was added to the main wash, as the contact time and wash temperature were the longest (30 min) and/or the highest (40 °C). However, this was not the case, as the results show that the colour difference is randomly distributed between the different wash cycles of prewash, main wash and rinse. Accordingly, it can be concluded that the addition of 3% HP to the liquid detergent during a household wash does not cause discoloration in most modern textiles. This can be explained by the fact that modern manufacturers of dyes and auxiliaries have adjusted to the fact that textiles are exposed to various oxidising agents in household laundry. Therefore, the dyes produced and used are resistant to oxidation by hydrogen peroxide [37].

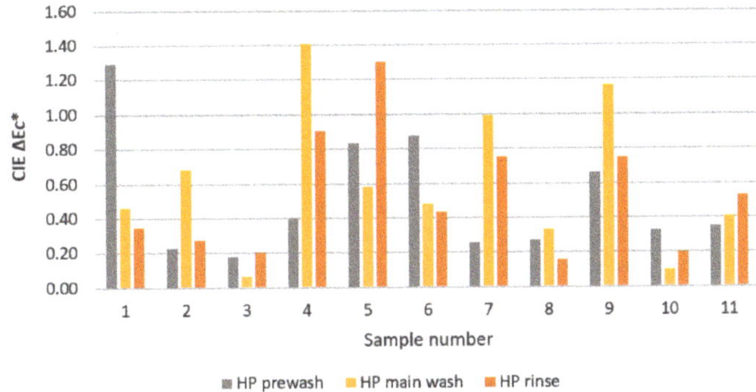

Figure 6. Colour difference, ΔE_c^*, of textile samples determined as the difference of CIE L*a*b* values of the samples after five consecutive washings without and with the addition of 3% HP in the prewash, main wash and rinse cycles.

2.4. Tensile Strength of Coloured Textile Samples

Figure 7 shows the tensile strength of textile samples washed with an addition of 3% HP at different washing cycles. In the statistical analysis, we used a paired *t*-test, bilateral and unilateral, and found that the tensile strength of the samples washed without hydrogen peroxide was not statistically significantly higher than that of the samples washed without peroxide at 95% confidence level. Thus, washing with 3% HP at any cycle of household laundering has no effect on the tensile strength of the washed textiles. At a temperature as low as 40 °C and a relatively short exposure time as used in household washing, hydrogen peroxide does not damage the textile fibres despite its high concentration.

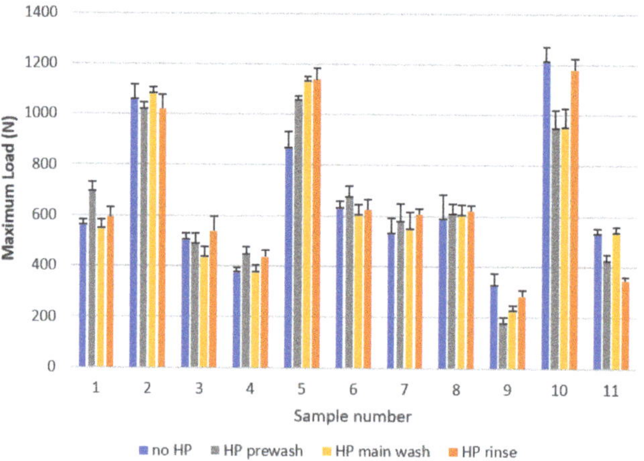

Figure 7. Tensile strength of textile samples after five consecutive washings without and with the addition of 3% HP in the prewash, main wash and rinse cycles at pH 7.

3. Materials and Methods

3.1. Materials

To test the effectiveness of 3% HP in soil removal, four different EMPA standard soil fabrics were prepared according to the standard SIST EN 60456:2005: Empa 101, Empa 114, Empa 116 and Empa 160, manufactured by Materials Research Products, UK. The properties of the Empa standard soil fabrics used are listed in Table 1.

Table 1. Code and properties of used Empa standard soiled fabrics.

Code	Description	Mass Per Unit Area (g/m^2)
Empa 101	100% Cotton soiled with carbon black/olive oil	90
Empa 114	100% Cotton soiled with red wine	200
Empa 116	100% Cotton soiled with blood/milk/ink	200
Empa 160	100% Cotton soiled with chocolate cream	200

To determine the effect of 3% HP on washed textiles, eleven fabrics (20 × 20 cm) of different material compositions, constructions and colours were prepared. The selected samples are shown in Figure 8, while their composition is shown in Table 2.

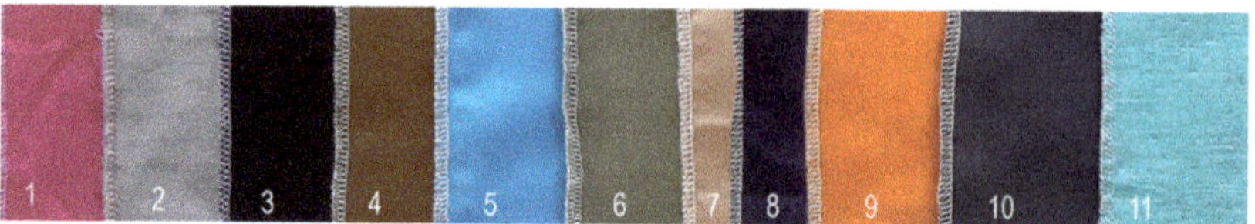

Figure 8. Textile samples of different material composition.

Table 2. Fabric material compositions.

Sample No.	Component 1	Component 2
1	Cotton	-
2	Cotton	-
3	Viscose	Linen
4	Cotton	Viscose, Linen
5	Cotton	Polyester
6	Viscose	Polyester
7	Cotton	Polyester
8	Wool	Polyester
9	Wool	Viscose, CA
10	Polyester	-
11	Polyester	-

3.2. Washing Procedure

Washing was performed according to the standard SIST EN 60456:2005. The standard soiled fabrics and the coloured fabric samples were washed separately in a Gorenje household washing machine, model Asko (Gorenje, Slovenia). Household washing was performed at 40 °C using a liquid detergent Ox (Yuco hemija d.o.o., Bački Jarak, Serbia). The ballast load consisted of 100% cotton fabrics (170–190 g/m^2) with a total weight of 2.45 kg. An appropriate amount of hydrogen peroxide in the form of 35% solution (Belox, Belinka Perkemija d.o.o., Ljubljana-Črnuče, Slovenia) was added to the washing machine in different washing cycles. Separate washing cycles were carried out for determination of disinfection activity, soil removal and colour/tensile strength properties.

The "Universal with prewash" programme was used for all wash cycles. The conditions of the wash cycles are listed in Table 3.

Table 3. Washing cycles conditions.

Wash Cycle	Time (min)	Wash Bath Volume (l)	Temperature (°C)
Prewash	16	3	28
Main Wash	30	10	40
Rinse	10	3	26

Four different washing procedures were preformed:

1. No HP: washing with liquid detergent only
2. HP prewash: 3% H_2O_2 in prewash + wash with liquid detergent
3. HP main wash: washing with liquid detergent + 3% H_2O_2 in the main wash
4. HP rinse: washing with liquid detergent + 3% H_2O_2 in a rinse wash

For the determination of disinfection efficacy and the determination of soil removal, one washing cycle was performed. For the determination of colour/tensile properties, each procedure was repeated 5 times for each batch of samples. Between each washing procedure, the samples were air dried. The pH of all washing baths was a value of 7.

3.3. Analysis

Disinfection efficacy was tested according to the standard EN 16616:2015. In brief: Standard bacterial strains of cultures *E. coli* ATCC 35218, *P. aeruginosa* ATCC 27853 and *S. aureus* ATCC 25923 were taken from the bacterial collection of the Faculty of Health Sciences. The bacteria from the collection were transferred to nutrient agar and incubated at 37 °C for 24 h. Then, a single colony of the strain was transferred from nutrient agar to nutrient broth (Biolife, Milan, Italy) and incubated under the same conditions. For recovery, the bacterial suspension was centrifuged at 3000× g for 15 min. The liquid was removed, and the pellet was mixed with an appropriate amount of 0.9% NaCl to obtain a cell concentration of 1.5×10^9 CFU mL^{-1} (OD 620 nm). The suspension was then centrifuged again and resuspended with defibrinated sheep blood in the same volume.

Cotton carriers were cut into 1 cm^2, boiled in distilled water and autoclaved. The cotton carriers were soaked in the bacterial suspension in defibrinated sheep blood for 15 min at room temperature. After drying at room temperature in a safety cabinet, the carriers were transferred to sterile cotton bags and the disinfection process was carried out in the washing machine. After filling the washing machine with ballast and cotton carriers, 12.5 mL of defibrinated sheep blood per kg of ballast and 5 g/L of liquid detergent were added. Before each wash cycle (i.e., prewash, main wash and rinse), an appropriate amount of 35% HP was added to the washer to achieve a concentration of 3%. At the end of the disinfection step, the bags containing the carriers were removed. Each sample carrier (Ns) was transferred to a separate tube containing a 5 mL neutraliser (0.25 mol/L phosphate buffer) and glass beads. Bacterial cells were detached from the carriers by vortex shaking for 10 min. Serial dilution of the sample liquid was then performed, and 1 mL was transferred to a Petri dish and covered with agar. Endo agar (Merck) was used to obtain *E. coli*, cetrimide agar (Sigma-Aldrich, St. Louis, MO, USA) was used for *P. aeruginosa* and mannitol salt agar (Biolife) was used for *S. aureus*. Agar plates were incubated at 37 °C for 24 h, colonies were counted and results were expressed as log CFU cm^{-2}. All experiments were performed with five parallel trials and three replicates. Cotton slides with bacteria not exposed to the washing and disinfection process were considered as control (Nc). The reduction of bacterial cells was expressed as relative reduction (R) in percentage and absolute reduction was calculated as log reduction RL = (Ns/Nc) (log CFU cm^{-2}).

To determine the influence of the presence of 3% HP in the studied washing cycles on soil removal and colour change of the tested textile samples, colour measurements were performed using a Datacolour Spectraflesh 600 PLUS-CT spectrophotometer operated with Datacolour Datamaster software with the following settings: Illuminant D65, large area view, specular excluded, UV included and 10° standard observer angle.

For soil removal efficiency determination, the difference in lightness between washed and unwashed standard soils, ΔLs*, was calculated as follows:

$$\Delta L_S^* = L_W^* - L_{UN}^*$$

(1)

where L_W^* and L_{UN}^* stands for CIE L* coordinate values of the washed and unwashed standard soils sample.

For colour change determination of the tested coloured textile samples, the colour difference (ΔE*) was calculated using the following equation:

$$\Delta E^* = \sqrt{(\Delta L^*)^2 + (\Delta a^*)^2 + (\Delta b^*)^2}$$

(2)

where ΔL*, Δa* and Δb* are differences between the lightness (L*), green-red (a*) and blue-yellow (b*) colour coordinates of the two samples, i.e., samples washed with washing agent in the presence of hydrogen peroxide and samples washed only with washing agent. Five measurements per sample were performed.

Measurements of tenacity at maximum load were performed by an Instron 5567 dynamometer in accordance with SIST ISO 5081:1996. Measurements were performed in the warp direction for ten fabric samples. Prior to testing, the samples were conditioned at 65 ± 2% relative humidity and 20 ± 1 °C for 24 h.

4. Conclusions

The introduction of an eco-friendly hydrogen peroxide into household low-temperature (i.e., 40 °C) washing processes can significantly increase the hygiene of the washed textiles and the hygiene of the internal parts of the washing machine by eliminating microorganisms without impairing the intrinsic properties of textiles. Such improved laundry hygiene is of significant importance in preventing the transition of pathogenic microorganisms from person to person through the contact of contaminated textiles during the washing process. Accordingly, the highest disinfection activity against the tested bacteria *E. coli*, *S. aureus* and *P. aeruginosa* was obtained by adding 3% HP to the main wash cycle, corresponding to a reduction > 6.6 log. Despite the relatively high concentration, the addition of 3% HP to the household wash cycle at 40 °C with a liquid detergent does not cause discoloration or damage to textiles, regardless of their raw composition. Moreover, the addition of 3% HP has a positive effect on the removal of standard soiling during different wash cycles, which was highest when added during the main wash. Therefore, it can be concluded that a small addition of eco-friendly HP into the main wash assures disinfection and improved soil removal efficiency of low-temperature washing while maintaining the intrinsic mechanical properties of the textile apparel.

Author Contributions: Investigation, writing—original draft preparation, review and editing, supervision, funding acquisition, P.F.T.; investigation, data curation, writing—review and editing, K.B.; investigation, data curation, writing—review and editing, R.F.; data curation, writing—review and editing, B.T. All authors have read and agreed to the published version of the manuscript.

Funding: This research was funded by the Slovenian Research Agency (Program P2-0213; Infrastructural Centre RIC UL-NTF; applied project L2-9199) and by the company Belinka Perkemija, Slovenia.

Institutional Review Board Statement: Not applicable.

Informed Consent Statement: Not applicable.

Data Availability Statement: Not applicable.

Conflicts of Interest: The authors declare no conflict of interest.

Sample Availability: Samples are not available from the authors.

References

1. Commission Regulation (EU) 2019/2023 of 1 October 2019 Laying Down Ecodesign Requirements for Household Washing Machines and Household Washer-Dryers Pursuant to Directive 2009/125/EC of the European Parliament and of the Council, Amending Commission Regulation (EC) No 1275/2008 and Repealing Commission Regulation (EU) No 1015/2010 (Text with EEA Relevance.) C/2019/2124. 2019. Available online: http://data.europa.eu/eli/reg/2019/2023/oj/ (accessed on 9 November 2021).
2. Pakula, C.; Stamminger, R. Electricity and water consumption for laundry washing by washing machine worldwide. *Energy Effic.* **2010**, *3*, 365–382. [CrossRef]
3. Pakula, C.; Stamminger, R. Energy and water savings potential in automatic laundry washing processes. *Energy Effic.* **2015**, *8*, 205–222. [CrossRef]
4. Laitala, K.; Boks, C.; Klepp, I.G. Potential for environmental improvements in laundering. *Int. J. Consum. Stud.* **2011**, *35*, 254–264. [CrossRef]
5. Ferri, A.; Osset, M.; Abeliotis, K.; Ambert, C.; Candan, C.; Owens, J.; Stamminger, R. Laundry performance: Effect of detergent and additives on consumer satisfaction. *Tenside Surfactants Deterg.* **2016**, *53*, 375–386. [CrossRef]
6. Lambert, E.; Bichler, S.; Stamminger, R. Hygiene in domestic laundering–Consumer behavior in Germany. *Tenside Surfactants Deterg.* **2015**, *52*, 441–446. [CrossRef]
7. Kruschwitz, A.; Karle, A.; Schmitz, A.; Stamminger, R. Consumer Laundry Practices in Germany. *Int. J. Consum. Stud.* **2014**, *38*, 265–277. [CrossRef]
8. Rogers, D. Making Laundry Cool. Washing Trends and Innovations. Consumers and Brand Owners Alike Want More Sustainable Laundry Practices to Become the Norm. How Is the Transition Being Achieved Effectively? Du Pont. 2015. Available online: http://fhc.biosciences.dupont.com/fileadmin/user_upload/live/fhc/Making_Laundry_Cool2.pdf/ (accessed on 9 November 2021).
9. Betz, M.; Cerny, G. Antimicrobial effects of bleaching agents Part 2–Studies on bacteria, yeast, mould and phage. *Tenside Surfactants Deterg.* **2001**, *38*, 242–249.
10. Abney, S.E.; Ijaz, M.K.; McKinney, J.; Gerba, C.P. Laundry hygiene and odour control: State of the science. *Appl. Environ. Microb.* **2021**, *87*, e300220. [CrossRef]
11. Bockmühl, D.P.; Schages, J.; Rehberg, L. Laundry and textile hygiene in healthcare and beyond. *Microb. Cell* **2019**, *6*, 299–306. [CrossRef]
12. Heudorf, U.; Gasteyer, S.; Muller, M.; Serra, N.; Westphal, T.; Reinheimer, C.; Kempf, V. Handling of laundry in nursing homes in Frankfurt am Main, Germany, 2016–laundry and professional clothing as potential pathways of bacterial transfer. *GMS Hyg. Infect. Control* **2017**, *12*, eDoc20. [CrossRef]
13. Chiereghin, A.; Felici, S.; Gibertoni, D.; Foschi, C.; Turello, G.; Piccirilli, G.; Gabrielli, L.; Clerici, P.; Landini, M.P.; Lazzarotto, T. Microbial contamination of medical staff clothing during patient care activities: Performance of decontamination of domestic versus industrial laundering procedures. *Curr. Microbiol.* **2020**, *77*, 1159–1166. [CrossRef]

14. Reynolds, K.A.; Verhougstraete, M.P.; Mena, K.D.; Sattar, S.A.; Scott, E.A.; Gerba, C.P. Quantifying pathogen infection risks from household laundry practices. *J. Appl. Microbiol.* **2021**, in press. [CrossRef]

15. Honisch, M.; Brands, B.; Weide, M.; Speckmann, H.D.; Stamminger, R.; Bockmuhl, D. Antimicrobial efficacy of laundry detergents with regard to time and temperature in domestic washing machines. *Tenside Surfactants Deterg.* **2016**, *53*, 547–552. [CrossRef]

16. Laitala, K.; Jensen, H.M. Cleaning effect of household laundry detergents at low temperatures. *Tenside Surfactants Deterg.* **2010**, *47*, 413–420. [CrossRef]

17. Pusic, T.; Jelicic, J.; Nuber, M.; Soijacic, I. Investigation of bleach active compounds in washing bath. *Tekstil* **2007**, *56*, 412–417.

18. Brands, B.; Brinkmann, A.; Bloomfield, S.; Bockmühl, D.P. Microbicidal action of heat, detergents and active oxygen bleach as components of laundry hygiene. *Tenside Surfactants Deterg.* **2016**, *53*, 495–501. [CrossRef]

19. Hickman, W.S. Peracetic acid and its uses in fibre bleaching. *Rev. Prog. Color.* **2002**, *32*, 13–27. [CrossRef]

20. Yun, C.S.; Patwary, S.; LeHew, M.L.A.; Kim, J. Sustainable care of textile products and its environmental impact: Tumble-drying and ironing processes. *Fiber. Polym.* **2017**, *18*, 590–596. [CrossRef]

21. Beck, R.H.F.; Koch, H.; Mentech, J. Bleach Activators. In *Carbohydrates as Organic Raw Materials III*; Van Bekkum, H., Röper, H., Voragen, F., Eds.; VCH Weinhem: New York, NY, USA, 1996; pp. 295–306. [CrossRef]

22. Jacksch, S.; Kaiser, D.; Weis, S.; Weide, M.; Ratering, S.; Schnell, S.; Egert, M. Influence of sampling site and other environmental factors on the bacterial community composition of domestic washing machines. *Microorganisms* **2020**, *8*, 30. [CrossRef] [PubMed]

23. Jacksch, S.; Zohra, H.; Weide, M.; Schnell, S.; Egert, M. Cultivation-based quantification and identification of bacteria at two hygienic key sides of domestic washing machines. *Microorganisms* **2021**, *9*, 905. [CrossRef] [PubMed]

24. Neral, B.; Arnuš, S. Textile Washing Hygiene and the Covid-19 Pandemic. *Tekstilec* **2021**, *64*, SI 7–SI 17.

25. Köse, H.; Yapar, N. The comparison of various disinfectants' efficacy on Staphylococcus aureus and Pseudomonas aeruginosa biofilm layers. *Turk. J. Med. Sci.* **2017**, *47*, 1287–1294. [CrossRef] [PubMed]

26. Upson, S.; Clarke, C. *Support to the Evaluation of Regulation (EC) No 648/2004 (Detergents Regulation)*; European Commission: Brussels, Belgium, 2018; Available online: https://op.europa.eu/sl/publication-detail/-/publication/ad2fa114-e952-11e8-b690 -01aa75ed71a1/ (accessed on 9 November 2021).

27. Urban, M.V.; Rath, T.; Radtke, C. Hydrogen peroxide (H2O2): A review of its use in surgery. *Wien. Med. Wochenschr.* **2019**, *169*, 222–225. [CrossRef] [PubMed]

28. Fijan, S.; Sostar-Turk, S. Antimicrobial activity of selected disinfectants used in a low temperature laundering procedure for textiles. *Fibres Text. East. Eur.* **2010**, *18*, 89–92.

29. Altenbaher, B.; Šoštar Turk, S.; Fijan, S. Ecological parameters and disinfection effect of low-temperature laundering in hospitals in Slovenia. *J. Clean. Prod.* **2011**, *19*, 253–258. [CrossRef]

30. McLaren, K.; McCauley, E.; O'Neill, B.; Tinker, S.; Jenkins, N.; Sehulster, L. The efficacy of a simulated tunnel washer process on removal and destruction of Clostridioides difficile spores from health care textiles. *Am. J. Infect. Control.* **2019**, *47*, 1375–1381. [CrossRef]

31. Boateng, M.K.; Price, S.L.; Huddersman, K.D.; Walsh, S.E. Antimicrobial activities of hydrogen peroxide and its activation by a novel heterogeneous Fenton's-like modified PAN catalyst. *Appl. Environ. Microb.* **2011**, *111*, 1533–1543. [CrossRef]

32. Brul, S.; Coote, P. Preservative agents in foods: Mode of action and microbial resistance mechanisms. *Int. J. Food Microbiol.* **1999**, *50*, 1–17. [CrossRef]

33. Choi, H.; Chatterjee, P.; Lichtfouse, E.; Martel, J.A.; Hwang, M.; Jinadatha, C.; Sharma, V.K. Classical and alternative disinfection strategies to control the COVID-19 virus in healthcare facilities: A review. *Env. Chem. Lett.* **2021**, *19*, 1945–1951. [CrossRef]

34. Johansson, I.; Somasundaran, P. (Eds.) *Handbook for Cleaning/Decontamination of Surfaces*; Elsevier: Amsterdam, The Netherlands, 2007.

35. Hooker, E.A.; Ulrich, D.; Brooks, D. Successful Removal of Clostridioides Difficile Spores and Pathogenic Bacteria from a Launderable Barrier Using a Commercial Laundry Process. *Infect. Dis.* **2020**, *13*, 1178633720923657. [CrossRef]

36. Forte Tavčer, P. Influence of bleach activators in removing different soils from cotton fabric. *Fibres Text. East. Eur.* **2020**, *141*, 74–78. [CrossRef]

37. Phillips, D.; Scotney, J. Oxidative-bleach fading of dyed cellulosic textiles when washed in a detergent containing a bleach activator. *AATCC Rev.* **2002**, *2*, 50–53.

Article

Influence of Initial pH Value on the Adsorption of Reactive Black 5 Dye on Powdered Activated Carbon: Kinetics, Mechanisms, and Thermodynamics

Branka Vojnović [1], Mario Cetina [1,*], Petra Franjković [1] and Ana Sutlović [2]

1 Department of Applied Chemistry, Faculty of Textile Technology, University of Zagreb, Prilaz baruna Filipovića 28a, 10000 Zagreb, Croatia; branka.vojnovic@ttf.unizg.hr (B.V.); petra@franjkovic.com (P.F.)

2 Department of Textile Chemistry and Ecology, Faculty of Textile Technology, University of Zagreb, Savska c. 16/9, 10000 Zagreb, Croatia; ana.sutlovic@ttf.unizg.hr

* Correspondence: mario.cetina@ttf.unizg.hr

Abstract: The aim of this work was to investigate the influence of initial pH value (pH_0) on the isothermal adsorption of Reactive Black 5 (RB5) dye on commercial powdered activated carbon. Four initial pH values were chosen for this experiment: pH_0 = 2.00, 4.00, 8.00, and 10.00. In order to investigate the mechanism of adsorption kinetic, studies have been performed using pseudo-first-order and pseudo-second-order kinetic models as well as an intraparticle diffusion model. In addition, thermodynamic parameters of adsorption were determined for pH_0 = 4.00. Results of this research showed that the initial pH value significantly influences the adsorption of RB5 dye onto activated carbon. The highest adsorption capacities (q_e) and efficiencies of decolouration were observed for initial pH values of pH_0 = 2.00 (q_e = 246.0 mg g^{-1}) and 10.00 (q_e = 239.1 mg g^{-1}) due to strong electrostatic interactions and attractive $\pi\cdots\pi$ interactions, respectively. It was also shown that the adsorption of RB5 dye on activated carbon at all initial pH values is kinetically controlled, assuming a pseudo-second-order model, and that intraparticle diffusion is not the only process that influences on the adsorption rate.

Keywords: adsorption; activated carbon; Reactive Black 5; kinetics; thermodynamics

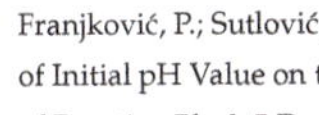
check for **updates**

Citation: Vojnović, B.; Cetina, M.; Franjković, P.; Sutlović, A. Influence of Initial pH Value on the Adsorption of Reactive Black 5 Dye on Powdered Activated Carbon: Kinetics, Mechanisms, and Thermodynamics. *Molecules* **2022**, *27*, 1349. https://doi.org/10.3390/molecules27041349

Academic Editor: Baljinder Kandola

Received: 30 December 2021
Accepted: 13 February 2022
Published: 16 February 2022

Publisher's Note: MDPI stays neutral with regard to jurisdictional claims in published maps and institutional affiliations.

1. Introduction

The textile industry has a positive impact on economic development around the world, because of constantly growing demands for textile products. The main environmental textile industry problem is the fact that its intensive development causes an increased use of resources, especially in terms of water consumption, and releases highly contaminated wastewaters with a wide range of unprocessed harmful and toxic chemicals. A further problem is that some less harmful chemicals are used in huge quantities and therefore also produce a lot of waste [1,2]. Wet treatment textile industry processes (e.g., finishing, dyeing, printing etc.) are the main sources of toxic emissions into the environment—water, air, and soil. Many textile-processing operations generate large amounts of pollutants and pose a threat to the environment if they are not appropriately treated. One of the most studied textile industry process in the context of wastewaters is dyeing. Even very low concentrations of dyes in wastewaters are noticeable, which makes water aesthetically and ecologically unacceptable. At the same time, colour reduces the transparency of the water and creates the impression of a high degree of pollution. The presence of water colouring causes less penetration of sunlight into the depths of the water, which leads to the disruption of the process of photosynthesis and prevents animals' orientation. In addition, some dyes are toxic, while auxiliaries such as carriers, metals, salts, etc. used in the dying process also contribute to the water pollution. Therefore, the wastewater stream from the textile dyeing process besides unutilized dyes (about 8−20% of the total pollution load due to incomplete dye exhaustion) contains other auxiliary chemicals. Textile dyeing wastewaters also contain high pH values, high salt concentrations, high total suspended solids, high temperatures, and high chemical oxygen demand (COD) and biochemical oxygen demand (BOD) values [3,4].

There are more than 100,000 commercially available dyes, and more than 700,000 tons of dyes are produced annually. As a result of their benefits, such as vibrant colours, high colourfastness, and simplicity of application, reactive dyes are one of the most often used dyes [4]. However, the removal of reactive dyes from wastewaters is generally difficult. They are usually resistant to aerobic digestion and stable to light, heat, and oxidising agents. Due to their relatively poor biodegradation in aerobic circumstances (particularly those containing azo groups) and their increased use, they had an impact on traditional methods for treating textile wastewaters [4,5]. Therefore, finding the appropriate method for the treatment of a particular type of textile wastewater is very important. For this purpose, wastewater is collected and subjected to physical, chemical, and/or biological treatment processes before being returned to the environment.

Many different methods and techniques have been reported for the removal and/or decolourisation of high soluble reactive dyes from wastewaters: coagulation and flocculation, chemical oxidation, biological treatment, membrane separation, reverse osmosis, etc. [4,6–9]. Chemical and biological methods could be very effective, but they require specialised equipment; they are energy intensive, use an excess of chemicals, and may produce large amounts of solid waste or could generate environmentally unsuitable by-products. The high solubility of reactive dyes causes a relatively poor colour removal efficiency by the coagulation and flocculation method, so that they have a limited application [7]. On the other hand, physical and physicochemical methods (adsorption, ion exchange, membrane filtration, etc.) have proven to be very effective for the removal of reactive dyes. Adsorption is one of the best treatment methods due to its flexibility, simplicity of design, industrial application, and sensitivity to toxic pollutants [4,8,10]. Many studies on the development and practical application of different effective adsorbents for reactive dyes removal, including Reactive Black 5 dye discussed in this paper, have been reported. These studies and recent reviews include adsorbents such as natural and by-product adsorbents [11–13], surface-modified natural adsorbents [14], nanostructures [15,16], chitosan-based adsorbents [10,17,18], aminosilosan functionalised silica [19], and mixed silica-alumina oxides [20,21]. Modern adsorbents, such as nanostructures, are highly effective for Reactive Black 5 removal, but they are expensive, and therefore, it is essential to regenerate and reuse them for the water treatment [15,16]. The advantages of natural and by-product adsorbents is their low cost, easy accessibility, and usage of waste materials such as agricultural crop or industrial residues as a source for activated carbon production [11–13,22]. Activated carbon is one of the most popular adsorbents used for wastewater treatment due to its adsorption efficiency, great capacity, and very high quality of purified water [22–24]. In addition, adsorption on activated carbon generally does not result in the formation of harmful substances, and it can be used for the control of precursors that may form toxic compounds [25].

It is known that the adsorption process significantly depends on several factors, including temperature, optimal dose of adsorbent, contact time, initial dye concentration, and pH value. Therefore, in this study, the adsorption of Reactive Black 5 (RB5) onto commercial powdered activated carbon at different initial pH values was examined. Most adsorption studies have been conducted on acidic and neutral media [10]. On the other hand, we performed adsorption experiments in a wider range of initial pH values, from $pH_0 = 2.00$ to $pH_0 = 10.00$. It should be pointed out that the adsorption of RB5 dye on commercial powdered activated carbon at high pH values, such as $pH_0 = 10.00$, which is used in real RB5 dye industrial application (e.g., for cellulose fibres dying), is rarely described in the literature. We monitored pH value during the whole adsorption time, which helped us to presume adsorption mechanisms at different initial pH values. The kinetics and thermodynamics for this adsorption system were also investigated.

2. Results and Discussion

2.1. Effect of Initial pH Value on Adsorption Process

The objective of this work was to evaluate the influence of initial pH value (pH_0) on the treatment of dye-rich textile wastewaters by the adsorption process. For this study, we chose an RB5 dye concentration of $c_0 = 500$ mg dm^{-3}, which is a possible dye concentration in textile wastewaters. The amounts of adsorbed dye at time t (q_t) for different initial pH values are given in Table 1, while the liquid phase dye concentrations (c_t) during adsorption process are presented in Figure 1.

Table 1. Amount of adsorbed Reactive Black 5 (RB5) dye (q_t) at time t for different initial pH values.

$t/$ min	q_t/mg g^{-1}			
	$pH_0 = 2.00$	$pH_0 = 4.00$	$pH_0 = 8.00$	$pH_0 = 10.00$
15	117.1	65.5	70.5	111.0
30	132.0	86.0	84.7	116.4
60	155.9	98.5	99.0	126.7
120	177.9	115.8	128.5	161.7
960	246.0	193.0	194.9	239.1

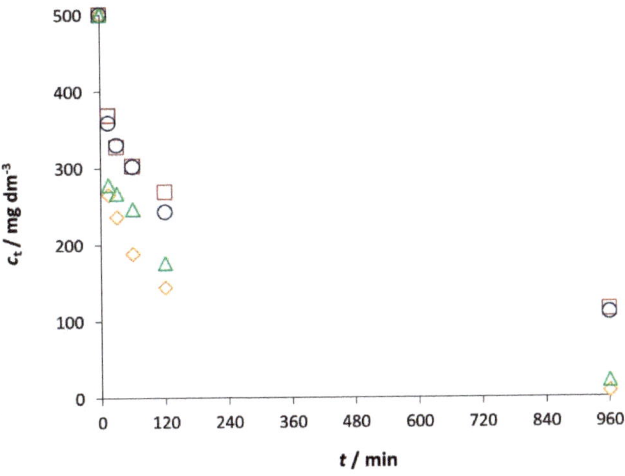

Figure 1. Effect of initial pH values on the RB5 dye concentration in the liquid phase (c_t) after appropriate time of adsorption (t) ($\Diamond$ $pH_0 = 2.00$; $\Box$ $pH_0 = 4.00$; O $pH_0 = 8.00$; $\triangle$ $pH_0 = 10.00$).

The data in Table 1 show that the initial pH value significantly influences adsorption capacity, i.e., the amount of adsorbed dye at equilibrium (q_e). The highest adsorption capacities were observed for initial pH values of $pH_0 = 2.00$ and 10.00 ($q_e = 246.0$ and 239.1 mg g^{-1}, respectively), while values for $pH_0 = 4.00$ and 8.00 are much lower ($q_e < 200$ mg g^{-1}). From Figure 1, it is obvious that dye concentration in the liquid phase decreases very fast. The plots can be approximately divided into several regions: very fast initial adsorption, then a milder and gradual decrease of dye concentration, which then reaches the equilibrium state. In this figure, one can also see that dye concentrations in the liquid phase for $pH_0 = 4.00$ and 8.00 are approximately equal for all contact times, with the exception for a contact time of 2 h.

To calculate the efficiency of decolouration (E_d) for all initial pH values, we compared with results obtained using the same procedure of adsorption experiment and applying the same conditions (dye concentration, mass of adsorbent, and temperature) but without changing the pH value before adsorption [26] (Table 2).

Table 2. Efficiency of decolouration (E_d) after appropriate time of adsorption (t) for different initial pH_0 values.

$t/$ min	$E_d/\%$				
	$pH_0 = 2.00$	$pH_0 = 4.00$	$pH_0 = 8.00$	$pH_0 = 10.00$	$pH_0 = 4.83$ [26]
15	46.8	26.2	28.2	44.4	– [a]
30	52.8	34.4	33.9	46.5	33.4
60	62.3	39.4	39.6	50.7	40.3
120	71.1	46.3	51.4	64.7	52.8
960	98.4	77.2	78.0	95.6	76.3

[a] adsorption experiment was not performed.

As already noticed for adsorption capacities, the results show that the biggest values of E_d were observed for $pH_0 = 2.00$ and $pH_0 = 10.00$, which at equilibrium amount to 98.4% and 95.6%, respectively. After 15 min, E_d values already reached ca. 45% and are almost twice as big as those for $pH_0 = 4.00$ and $pH_0 = 8.00$. At all time intervals, the E_d values follow the sequence: E_d ($pH_0 = 2.00$) > E_d ($pH_0 = 10.00$) > E_d ($pH_0 = 4.00 \approx pH_0 = 8.00$). In addition, the values obtained for $pH_0 = 4.00$ are approximately equal to those for $pH_0 = 4.83$, when the pH value before the adsorption experiment was not changed [26]. The fact that E_d values are the biggest for $pH_0 = 2.00$ can be explained by the activated carbon surface charge and interactions present between activated carbon and RB5 dye, which dissociate on coloured anion and sodium ions. If the pH value is less than the activated carbon point of zero charge (pH_{PZC}), its surface is positively charged and adsorbs the coloured anions due to strong electrostatic attraction. The lower the pH value relative to the pH_{PZC}, the stronger these attractions. It has been previously reported that the pH_{PZC} values for commercial powdered activated carbons range from 6.50 to 7.33 [27–30]. Such values support the fact that extremely strong adsorption at $pH_0 = 2.00$ could be attributed to very strong electrostatic interactions between the surface of the activated carbon and dye anions. The protonated groups of activated carbon are mainly carboxylic, hydroxyl, and chromenic. An example of such interaction between positively charged protonated hydroxyl groups of activated carbon and negatively charged sulfonic groups of RB5 is shown in Scheme 1 (interaction a).

Throughout the whole adsorption time, pH values of dye solutions for $pH_0 = 2.00$ were practically identical to the initial pH values (Figure 2). In contrast, for $pH_0 = 4.00$, besides RB5 dye, the surface of activated carbon adsorbed more H^+ ions than OH^- ions [27], and therefore, the pH value after adsorption increased already after 15 min; then, it gradually formed a plateau and stabilised at pH $\approx$ 6.5. For an initial pH value of $pH_0 = 8.00$, the pH value after adsorption slightly decreased and stabilised at pH $\approx$ 7.4.

This indicates that electrostatic interactions at $pH_0 = 4.00$ and $pH_0 = 8.00$ are not the driving force of adsorption anymore. Therefore, it can be concluded that dye is most likely bound to the adsorbent by weaker interactions, hydrogen bonds, or/and van der Waals forces, which resulted in significantly smaller E_d values. Activated carbon carboxyl and phenolic groups may be responsible for the formation of hydrogen bonds with RB5 dye donors or acceptors groups, e.g., $-NH_2$, $-S=O$, and $-O-H$ [27]. Two examples, in which activated carbon and RB5 are donors and acceptors of hydrogen bonds, are presented in Scheme 1 (interactions b and c). The surface of activated carbon at $pH_0 = 10$ is negatively charged, which can lead to possible rejection between the activated carbon surface and RB5 dye anion. Then, high E_d values at this pH initial value could be attributed to attractive $\pi\cdots\pi$ interactions [31]. The structure of the RB5 dye (Figure 3) contains four aromatic rings that allow its binding to the aromatic graphene layers of activated carbon by $\pi \cdots \pi$ aromatic interactions [32,33]. An example of such an interaction between the naphthalene ring of RB5 and graphene layer of activated carbon is shown in Scheme 1 (interaction d). These interactions, established between two parallel and mutually shifted aromatic systems at a distance of ca. 3.5 Å and with the interaction energy up to 50 kJ mol^{-1} [34], can also form RB5 multilayers on the surface of the activated carbon. Despite strong adsorption at $pH_0 = 10$ observed in this study, it should be noted that the adsorption efficiency in most studies of RB5 adsorption on commercial powdered-activated carbon was still better in acidic media. The highest adsorption capacities were observed at pH values 2–4, which then decrease when the pH value increases to pH 6–7 [35–38].

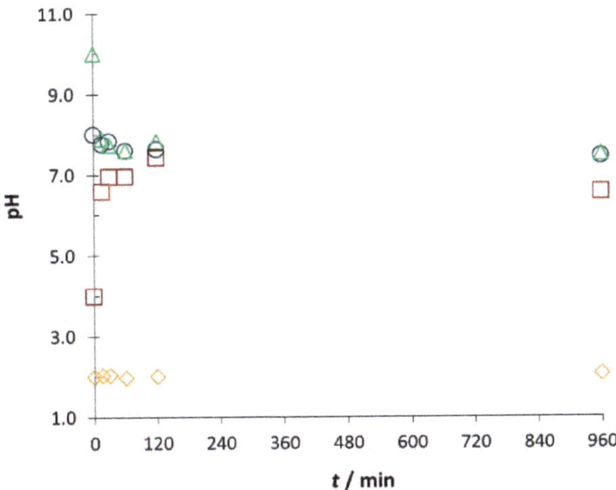

Scheme 1. Schematic representation of some interactions between activated carbon (AC) and RB5 anion: (**a**) electrostatic interaction between positively charged protonated hydroxyl group of AC and negatively charged sulfonic RB5 group, (**b**) O–H$\cdots$O hydrogen bonding between carboxyl group of AC and sulfonyl group of RB5, (**c**) N–H$\cdots$O hydrogen bonding between amino group of RB5 and carbonyl group of AC, (**d**) $\pi \cdots \pi$ interaction between naphthalene ring of RB5 and graphene layers of AC.

Figure 2. pH profiles of RB5 dye after adsorption for four initial pH values after appropriate time of adsorption (*t*) ($\Diamond$ $pH_0 = 2.00$; $\square$ $pH_0 = 4.00$; **O** $pH_0 = 8.00$; $\triangle$ $pH_0 = 10.00$.

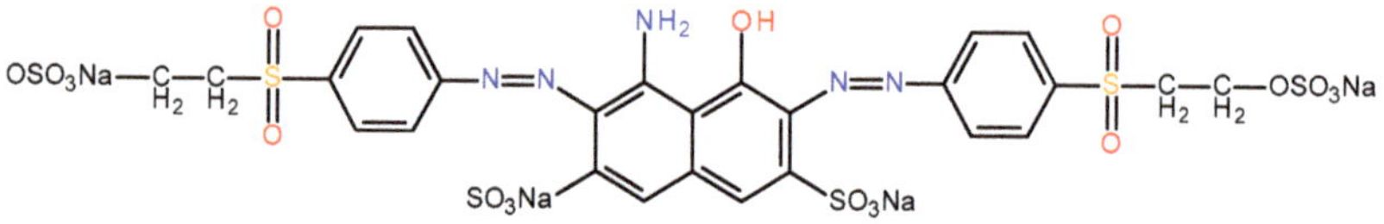

Figure 3. Chemical structure of Reactive Black 5 dye.

2.2. Kinetics of Adsorption

In order to investigate the mechanism of adsorption, kinetic studies have been performed. Kinetic study is important to an adsorption process because it depicts the uptake rate of the adsorbate and controls the residual time of the whole adsorption process. The experimental data were analysed by three kinetic models: pseudo-first-order, pseudo-second-order, and intraparticle diffusion models. The pseudo-first-order and pseudo-second-order are the most often used models for the determination of kinetic parameters.

2.2.1. Pseudo-First-Order and Pseudo-Second-Order Kinetic Models

Lagergren [39] proposed a rate equation for the sorption of solute from a liquid solution based on the solid capacity. The kinetic model of this rate equation is expressed by following equation:

$$\frac{dq_t}{dt} = k_1(q_e - q_t),$$
(1)

where k_1 is the rate constant of pseudo-first-order (min^{-1}).

Integrating this equation for the boundary conditions $t = 0$ to $t = t$ and $q_t = q_t$ gives a linear relationship expressed by the following equation:

$$\ln(q_e - q_t) = \ln q_e - k_1 \cdot t.$$
(2)

The pseudo-first-order kinetic constant k_1 can be determined by plotting $\ln(q_e - q_t)$ vs. time (t), and if the pseudo-first-order equation is applicable, the plot should give a linear relationship with a high value of correlation coefficient (R^2). The rate constant of pseudo-first-order (k_1) can be calculated from the slope of this plot when the amount of adsorbed dye is at equilibrium ($q_{e,calc.}$) from the intercept.

Ho and McKay [40,41] developed a second-order equation based on adsorption capacity. This kinetic model is given by the following equation:

$$\frac{dq_t}{dt} = k_2(q_e - q_t),$$
(3)

where k_2 is the rate constant of pseudo-second-order ($g\ mg^{-1}\ min^{-1}$).

Integrating this equation for the same boundary conditions as for the first-order gives the following equation in the linear form:

$$\frac{t}{q_t} = \frac{1}{k_2 \cdot q_e^2} + \frac{1}{q_e} \cdot t.$$
(4)

If the pseudo-second-order equation is applicable, the plot of t/q_t against time (t) should give a linear relationship, and it allows the calculation of amount of adsorbed dye at equilibrium ($q_{e,calc.}$) from the slope and afterwards the rate constant of pseudo-second-order (k_2) from the intercept. According to the pseudo-second-order model, as time approaches zero, the initial adsorption rate h ($mg\ g^{-1}\ min^{-1}$) can be also calculated using the following equation [41,42]:

$$h = k_2 \cdot q_{e,calc.}^2.$$
(5)

The kinetic parameters for pseudo-first-order and pseudo-second-order models for all four initial pH values are given in Table 3.

Table 3. Kinetic parameters for RB5 dye adsorption on activated carbon for different initial pH values.

pH$_0$	$q_{e,exp.}/$ mg g^{-1}	Pseudo-First-Order Model			Pseudo-Second-Order Model			
		$q_{e,calc.}/$ mg g^{-1}	R^2	$k_1/$ min^{-1}	$q_{e,calc.}/$ mg g^{-1}	R^2	$k_2/$ g mg^{-1}min^{-1}	$h/$ mg g^{-1} min^{-1}
2.00	246.0	136.6	0.9798	0.0060	256.4	0.9990	$1.169 \cdot 10^{-4}$	7.69
4.00	193.0	127.9	0.9388	0.0044	204.1	0.9968	$8.745 \cdot 10^{-5}$	3.64
8.00	194.9	134.3	0.9962	0.0058	204.1	0.9982	$9.656 \cdot 10^{-5}$	4.02
10.00	239.1	142.2	0.9696	0.0049	250.0	0.9979	$9.222 \cdot 10^{-5}$	5.76

The data in Table 3 show that values of the correlation coefficients (R^2) for the pseudo-first-order model obtained from the linear plots defined by Equation (2) are relatively high, from 0.939 to 0.996. However, there is a great disagreement between the experimental ($q_{e,exp.}$) and calculated ($q_{e,calc.}$) values of the amount of adsorbed dye at equilibrium. This suggests that this adsorption system is not a first-order reaction and that possibly the pseudo-second-order model provides better correlation of the data.

Based on the pseudo-second-order model, the calculated q_e values ($q_{e,calc.}$) for all pH$_0$ show much better agreement with experimental equilibrium values ($q_{e,exp.}$), while the values of the correlation coefficients are higher than 0.997 (Table 3). Therefore, it can be concluded that adsorption of RB5 dye on commercial activated carbon is kinetically controlled, assuming a pseudo-second-order rather than a pseudo-first-order process. As expected, a maximum k_2 value was obtained for the pH initial value of pH$_0$ = 2.00. As in the case of efficiency of decolouration (E_d) values, the initial adsorption rate follows the same sequence: h (pH$_0$ = 2.00) > h (pH$_0$ = 10.00) > h (pH$_0$ = 4.00 $\approx$ pH$_0$ = 8.00).

2.2.2. Intraparticle Diffusion Model

For evaluation of the diffusion mechanism, we also used the intraparticle diffusion model. Most adsorption processes involve three steps:

(i)   Mass transfer of adsorbate from the solution to adsorbent surface,
(ii)   Adsorption of adsorbate at a site on the surface of the adsorbent, and
(iii)   Intraparticle diffusion of the adsorbate in the pores of adsorbent and adsorption at the site.

Step (ii) is often assumed to be very fast, and therefore, it cannot be treated as a rate-limiting step, while the adsorption of large molecules, for which longer contact time is needed to reach equilibrium, is almost always considered to be diffusion controlled by external film resistance and/or internal diffusion mass transport or intraparticle diffusion [27].

Theoretical treatments of intraparticle diffusion yield complex mathematical relationships, which differ in form as functions of the geometry of the adsorbent particle, and the intraparticle diffusion model could be based on the following equation [27,42]:

$$q_t = k_i \cdot t^{0.5}, \tag{6}$$

where k_i is the intraparticle diffusion rate constant (mg g^{-1} min$^{-0.5}$).

If the intraparticle diffusion is a rate-limiting step of adsorption, i.e., if intraparticle diffusion controls the rate of adsorption, then plot q_t vs. $t^{0.5}$ should be linear and pass through the origin. If the plot presents multi-linearity, this indicates that intraparticle diffusion is not the only rate-controlling step and that two or more rate-controlling steps occur in the adsorption process [27,43]. Figure 4 shows the root time plots for the adsorption of RB5 onto commercial activated carbon for all initial pH values. All plots on this figure are not linear, i.e., they exhibit multi-linearity. Therefore, it could be concluded that intraparticle diffusion is not the only process that influences the adsorption rate and that multiple steps took place during the adsorption process.

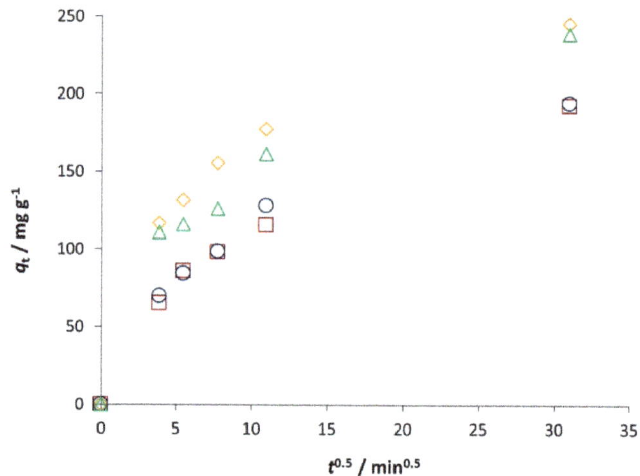

Figure 4. Root time plot for the adsorption of RB5 on activated carbon for four initial pH values ($\diamond$ pH$_0$ = 2.00; $\square$ pH$_0$ = 4.00; **O** pH$_0$ = 8.00; $\triangle$ pH$_0$ = 10.00).

2.3. Adsorption Thermodynamics

Thermodynamic adsorption parameters were calculated for the RB5 dye solution of initial pH value pH$_0$ = 4. This pH value was chosen because it was the closest to the pH$_0$ value of RB5 dye of the same concentration (pH$_0$ = 4.83 [26]). Standard Gibbs free energy change values ($\Delta G°$, kJ mol^{-1}) of the adsorption process can be calculated from the equation:

$$\Delta G° = -R\,T\,\ln\left(K_c\right),\tag{7}$$

where R is the universal gas constant, and T is temperature. K_c is the equilibrium constant calculated from the initial dye concentration (c_0, mg dm^{-3}) and the concentration of dye in the liquid phase at equilibrium (c_e, mg dm^{-3}) according to following equation [44]:

$$K_c = \frac{c_0 - c_e}{c_e}.\tag{8}$$

Then, the values of the two other thermodynamic parameters were calculated from the van't Hoff equation:

$$\ln\left(K_c\right) = -\frac{\Delta H°}{R}\cdot\frac{1}{T} + \frac{\Delta S°}{R},\tag{9}$$

where standard enthalpy change ($\Delta H°$) can be calculated from the slope, and standard entropy change ($\Delta S°$) can be calculated from the intercept of the plot $\ln(K_c)$ vs. $1/T$. The J. H. van't Hoff plot resulted in a straight line with a correlation coefficient of 92.9% (Figure 5).

Table 4 presents the thermodynamic parameters ($\Delta G°$, $\Delta H°$, $\Delta S°$) calculated from the experimental data by using Equations (7)–(9). The negative values of $\Delta G°$ show that the adsorption of RB5 onto commercial activated carbon was a spontaneous process whereby no energy input from outside of the system was required. As the higher negative value reflects a more energetically favourable adsorption, it can be concluded that adsorption at 328 K is energetically the most favourable. The positive value of $\Delta H°$ indicates the endothermic nature of adsorption process. As the adsorption process is usually exothermic, this phenomenon can be explained by the desorption of water molecules previously adsorbed on the dye molecule and the adsorption of dye molecules on the surface of the activated carbon [44].

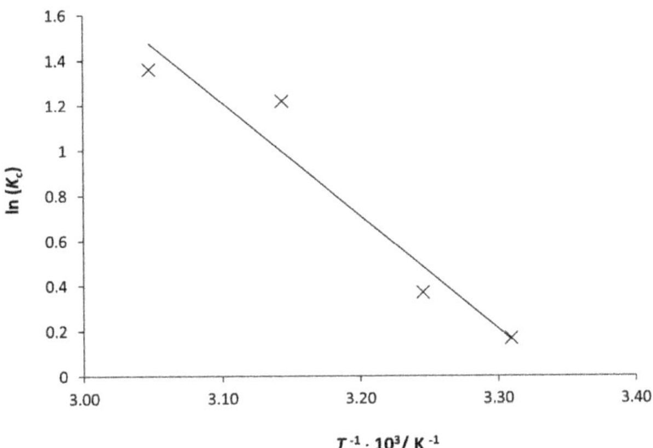

Figure 5. J. H. van't Hoff plot for adsorption of RB5 dye on activated carbon at $pH_0 = 4.00$.

Table 4. Thermodynamic parameters for the adsorption of RB5 dye on activated carbon for $pH_0 = 4.00$.

$T/$ K	$c_e/$ mg dm^3	$c_0 - c_e/$ mg dm^{-3}	K_c	$\Delta G°/$ kJ mol^{-1}	$\Delta H°/$ kJ mol^{-1}	$\Delta S°/$ J mol^{-1} K^{-1}	$T \cdot \Delta S°/$ kJ mol^{-1}
302	229.3	270.7	1.1805	−0.42			42.01
308	204.4	295.6	1.4462	−0.95	41.63	139.12	42.85
318	114.1	385.9	3.3821	−3.22			44.24
328	102.1	397.9	3.8972	−3.71			45.64

Based on the values of standard enthalpy change, the adsorption mechanism can be concluded, i.e., the type of bond between adsorbent and adsorbate. Energies of moderate-strength hydrogen bonds, such as those in DNA/RNA, acids, and alcohols range from 16 to 60 kJ mol^{-1} [34]. Therefore, the $\Delta H°$ value for this adsorption system of ca. 42 kJ mol^{-1} indicates hydrogen bonds as a driving force for adsorption, which is in accordance with the previous discussion of adsorption at $pH_0 = 4.00$ (Section 2.1). The positive value of $\Delta S°$ suggests the affinity of activated carbon for RB5 and increased randomness at the solid/liquid interface. During adsorption, the water coordinated molecules are displaced by dye molecules and consequently gain more translational entropy than is lost by dye molecules [29,43,44]. The data given in Table 4 also show an increase in $T \cdot \Delta S°$ values with increasing temperature, and that $T \cdot \Delta S°$ values are slightly bigger than the $\Delta H°$ value. This reveals that the adsorption process is a bit more dominated by entropic than enthalpic changes [45].

3. Materials and Methods

3.1. Chemicals

Reactive Black 5 (RB5) dye (Everzol Black B, supplied by Everlight Chemical Industrial Corp., Pineville, NC, USA, C.I. 20505, chemical formula: $C_{26}H_{21}N_5Na_4O_{19}S_6$, $M_r = 991.82$; purity = 75–80%, w/w) was used for the adsorption experiment. The chemical structure of the dye was prepared by the Biovia Draw program, Aachen, Germany. Powdered activated carbon was purchased from Kemika company, Croatia (particles size: <40 μm 85%, >80 μm 5%). Adsorbent was dried in an oven at 105 °C for 24 h and stored in desiccator until it is used.

3.2. Batch Mode Adsorption Studies

Adsorption studies were conducted by contacting 50 mL of RB5 dye solution of concentration $c_0 = 500$ mg dm^{-3} at different initial pH values (pH$_0$ = 2.00, 4.00, 8.00, and 10.00 adjusted by micro-additions of hydrochloric acid and sodium hydroxide solutions, $c = 0.1$ mol dm^{-3}) with 0.1 g of powdered activated carbon in glass bottles. Suspensions were shaken at different contact times (15, 30, 60, 120 min, and 16 h until equilibrium is reached) with an impeller speed of 250 rpm at 45 ($\pm$1) °C (Heidolph Unimax 1010 with Incubator 1000). An adsorption time of 16 h was appropriate to reach adsorption equilibrium, as was determined in our previous study [23]. Adsorption experiments after 16 h were also performed at 29 ($\pm$1), 35 ($\pm$1), and 55 ($\pm$1) °C in order to determine the thermodynamic parameters of adsorption. All experiments were repeated three times under identical conditions to confirm the repeatability of the experiments. The experimental data in tables and points presented in figures are the average values of three repetitions. After agitation, suspensions were filtered through filter-paper blue ribbon, and the residual liquid-phase dye concentration after adsorption was determined spectrophotometrically by monitoring the absorbance using UV-Vis spectrophotometer (Lambda 20, Perkin Elmer, Cleveland, OH, USA) at a maximum absorbance wavelength ($\lambda_{\max} = 598$ nm). The calibration graph of absorbance versus concentration obeyed a linear Beer–Lambert relation. For all filtrates, pH values after adsorption were also measured.

The amount of adsorbed dye at time t, q_t (mg g^{-1}), and at equilibrium, q_e (mg g^{-1}), were calculated by using following equation:

$$q = \frac{V \cdot (c_0 - c_t)}{m}, \tag{10}$$

where c_0 is initial dye concentration ($c_0 = 500$ mg dm^{-3}), c_t is its concentration in the liquid phase at time t and at equilibrium ($t = 16$ h), V is the volume of liquid phase (dm^3), and m is mass of the adsorbent (g).

The efficiency of adsorption, i.e., efficiency of decolouration (E_d), is calculated by the following equation:

$$E_d = \frac{(c_0 - c_t)}{c_0} \times 100. \tag{11}$$

4. Conclusions

This study showed that the best adsorption results were obtained for initial pH values of pH$_0$ = 2.00 and 10.00, which was presumably due to electrostatic interactions and $\pi\cdots\pi$ interactions, respectively. On the other side, at pH$_0$ = 4.00 and 8.00, it can be assumed that RB5 dye is bound to the activated carbon by hydrogen bonds. This means that probably both chemical (electrostatic interactions) and physical adsorption ($\pi\cdots\pi$ interactions and hydrogen bonds) occurred depending on the initial pH values. Furthermore, it was shown that the adsorption of RB5 dye on commercial-activated carbon at all initial pH values is kinetically controlled, assuming a pseudo-second-order model, and that intraparticle diffusion is not the only process that influences the adsorption rate. Negative values of standard Gibbs free energy change indicate that the adsorption reaction is spontaneous in nature and that the adsorption of RB5 on activated carbon is energetically the most favourable at the highest temperature. A positive value of standard enthalpy change revealed the endothermic nature of the adsorption, while a positive value of standard entropy change suggests the increased randomness at the solid/liquid interface.

Author Contributions: Conceptualisation, M.C. and B.V.; methodology, M.C. and B.V.; formal analysis, P.F.; investigation, P.F.; resources, M.C. and A.S.; writing—original draft preparation, M.C. and B.V.; writing—review and editing, M.C. and A.S.; visualisation, A.S.; supervision, M.C. All authors have read and agreed to the published version of the manuscript.

Funding: The APC was funded by University of Zagreb scientists research grants for 2021 (TP22/21, PP/21, TP6/21 and TP15/21).

Institutional Review Board Statement: Not applicable.

Informed Consent Statement: Not applicable.

Data Availability Statement: Data are available in a publicly accessible repository.

Acknowledgments: This is a paper recommended by the 14th Scientific-Professional Symposium Textile Science and Economy, The University of Zagreb Faculty of Textile Technology.

Conflicts of Interest: The authors declare no conflict of interest.

Sample Availability: Samples of the compounds are available from the authors.

References

1. Blackburn, R.S. *Sustainable Textiles: Life Cycle and Environmental Impact*; Woodhead Publishing Limited: Abington Hall, UK, 2009.
2. Prasad, M.N.V.; Shih, K. *Environmental Materials and Waste*; Elsevier Academic Press: London, UK, 2016.
3. Tharakeswari, S.; Shabaridharan, K.; Saravanan, D. Textile Effluent Treatment Using Adsorbents. In *Handbook of Textile Effluent Remediation*; Yusuf, M., Ed.; Pan Stanford Publishing Pte: Singapore, 2018; pp. 116–146.
4. Ejder-Korucu, M.; Gürses, A.; Doğar, Ç.; Sharma, S.K.; Açikyildiz, M. Removal of organic dyes from industrial Effluents: An overview of physical and biotechnological applications. In *Green Chemistry for Dyes Removal from Waste Water: Research Trends and Applications*; John Wiley & Sons: Hoboken, NJ, USA, 2015; pp. 1–34.
5. Mahony, T.O.; Guibal, E.; Tobin, J. Reactive dye biosorption by Rhizopus arrhizus biomass. *Enzym. Microb. Technol.* **2002**, *31*, 456–463.
6. Ince, N.H.; Tezcanlı, G. Reactive dyestuff degradation by combined sonolysis and ozonation. *Dyes Pigments* **2001**, *49*, 145–153. [CrossRef]
7. Verma, A.K.; Dash, R.R.; Bhunia, P. Review on chemical coagulation/flocculation technologies for removal of colour from textile wastewaters. *J Environ. Manag.* **2012**, *93*, 154–168.
8. Holkar, C.R.; Jadhav, A.J.; Pinjari, D.V.; Mahamuni, N.M.; Pandit, A.B. A critical review on textile wastewater treatments: Possible approaches. *J. Environ. Manag.* **2016**, *182*, 351–366.
9. Bilinska, L.; Gmurek, M.; Ledakowicz, S. Application of Advanced Oxidation Technologies for Decolorization and Mineralization of Textile Wastewaters. *J. Adv. Oxid. Technol.* **2015**, *18*, 185–194. [CrossRef]
10. Jalali Sarvestani, M.R.; Doroudi, Z. Removal of Reactive Black 5 from Waste Waters by Adsorption: A Comprehensive Review. *J. Water Environ. Nanotechnol.* **2020**, *5*, 180–190.
11. Afsharnia, M.; Biglari, H.; Javid, A.; Zabihi, F. Removal of Reactive Black 5 dye from Aqueous Solutions by Adsorption onto Activated Carbon of Grape Seed. *Iran. J. Health Sci.* **2017**, *5*, 48–61. [CrossRef]
12. Aguiar, J.E.; Bezerra, B.T.C.; Siqueira, A.C.A.; Barrera, D.; Sapag, K.; Azevedo, D.C.S.; Silva, I.J. Improvement in the Adsorption of Anionic and Cationic Dyes from Aqueous Solutions: A Comparative Study using Aluminium Pillared Clays and Activated Carbon. *Sep. Sci. Technol.* **2014**, *49*, 741–751. [CrossRef]
13. Eren, Z.; Acar, F.N. Adsorption of Reactive Black 5 from an aqueous solution: Equilibrium and kinetic studies. *Desalination* **2006**, *194*, 1–10. [CrossRef]
14. Nabil, G.M.; El-Mallah, N.M.; Mahmoud, M.E. Enhanced decolorization of Reactive Black 5 dye by active carbon sorbent-immobilized-cationic surfactant (AC-CS). *J. Ind. Eng. Chem.* **2014**, *20*, 994–1002. [CrossRef]
15. De Luca, P.; Nagy, B.J. Treatment of Water Contaminated with Reactive Black-5 Dye by Carbon Nanotubes. *Materials* **2020**, *13*, 5508. [CrossRef]
16. Mengelizadeh, N.; Pourzamani, H. Adsorption of Reactive Black 5 Dye from Aqueous Solutions by Carbon Nanotubes and its Electrochemical Regeneration Process. *Health Scope* **2020**, *9*, e102443. [CrossRef]
17. Ali, I. New generation adsorbents for water treatment. *Chem. Rev.* **2012**, *112*, 5073–5091. [PubMed]
18. Vakili, M.; Zwain, H.M.; Mojiri, A.; Wang, W.; Gholami, F.; Gholami, Z.; Giwa, A.S.; Wang, B.; Cagnetta, G.; Salamatinia, B. Effective Adsorption of Reactive Black 5 onto Hybrid Hexadecylamine Impregnated Chitosan-Powdered Activated Carbon Beads. *Water* **2020**, *12*, 2242. [CrossRef]
19. Wawrzkiewicz, M.; Nowacka, M.; Klapiszewski, Ł.; Hubicki, Z. Treatment of wastewaters containing acid, reactive and direct dyes using aminosilosane functionalized silica. *Open Chem.* **2015**, *13*, 82–95. [CrossRef]
20. Wawrzkiewicz, M.; Wiśniewska, M.; Gun'ko, V.M. Application of silica–alumina oxides of different compositions for removal of C.I. Reactive Black 5 dye from wastewaters. *Adsorpt. Sci. Technol.* **2017**, *35*, 448–457.
21. Wawrzkiewicz, M.; Wiśniewska, M.; Gun'ko, V.M.; Zarko, V.I. Adsorptive removal of acid, reactive and direct dyes from aqueous solutions and wastewater using mixed silica–alumina oxide. *Powder Technol.* **2015**, *278*, 306–315. [CrossRef]
22. Ip, A.W.M.; Barford, J.P.; McKay, G. Reactive Black dye adsorption/desorption onto different adsorbents: Effect of salt, surface chemistry, pore size and surface area. *J. Colloid Interface Sci.* **2009**, *337*, 32–38. [CrossRef] [PubMed]
23. Bonić, I.; Palac, A.; Sutlović, A.; Vojnović, B.; Cetina, M. Removal of Reactive Black 5 dye from aqueous media using powdered activated carbon—Kinetics and mechanisms. *Tekstilec* **2020**, *63*, 151–161. [CrossRef]

24. Metcalf & Eddy Inc. Separation processes for Removal of Residual Constituents. In *Wastewater Engineering Treatment and Reuse*, 5th ed.; Metcalf & Eddy, Inc., Ed.; McGraw-Hill Education: New York, NY, USA, 2014; pp. 1224–1245.
25. Sulaymon, A.H.; Abood, W.M. Equilibrium and kinetic study of the adsorption of reactive blue, red, and yellow dyes onto activated carbon and barley husk. *Desalin. Water Treat.* **2014**, *52*, 5485–5493. [CrossRef]
26. Gaščić, A.; Sutlović, A.; Vojnović, B.; Cetina, M. Adsorption of reactive dye on activated carbon: Kinetic study and influence of initial dye concentration. In Proceedings of the 2nd International Conference the Holistic Approach to Environment, Sisak, Croatia, 28 May 2021; pp. 131–138.
27. Ip, A.W.M.; Barford, J.P.; McKay, G. A comparative study on the kinetics and mechanisms of removal of Reactive Black 5 by adsorption onto activated carbons and bone char. *Chem. Eng. J.* **2010**, *157*, 434–442. [CrossRef]
28. Machado, F.M.; Bergmann, C.P.; Fernandes, T.H.M.; Lima, E.C.; Royer, B.; Calvete, T.; Fagan, S.B. Adsorption of Reactive Red M-2BE dye from water solutions by multi-walled carbon nanotubes and activated carbon. *J. Hazard. Mater.* **2011**, *192*, 1122–1131. [PubMed]
29. Kumar, A.; Prasad, B.; Mishra, I.M. Adsorptive removal of acrylonitrile by commercial grade activated carbon: Kinetics, equilibrium and thermodynamics. *J. Hazard. Mater.* **2008**, *152*, 589–600. [PubMed]
30. Al-Degs, Y.; Khraisheh, M.A.M.; Allen, S.J.; Ahmad, M.N. Effect of carbon surface chemistry on the removal of Reactive dyes from textile effluent. *Water Res.* **2000**, *34*, 927–935. [CrossRef]
31. Hunter, C.A.; Sanders, J.K.M. The Nature of π-π Interactions. *J. Am. Chem. Soc.* **1990**, *112*, 5525–5534. [CrossRef]
32. Kyzas, G.Z.; Deliyanni, E.A.; Lazaridis, N.K. Magnetic modification of microporous carbon for dye adsorption. *J. Colloid Interface Sci.* **2014**, *430*, 166–173. [CrossRef]
33. Giannakoudakis, D.A.; Kyzas, G.Z.; Avranas, A.; Lazaridis, N.K. Multi-parametric adsorption effects of the reactive dye removal with commercial activated carbons. *J. Mol. Liq.* **2016**, *213*, 381–389. [CrossRef]
34. Steed, J.W.; Turner, D.R.; Wallace, K.J. *Core Concepts in Supramolecular Chemistry and Nanochemistry*; John Wiley & Sons, Ltd.: Chichester, UK, 2007.
35. Lee, J.-W.; Choi, S.-P.; Thiruvenkatachari, R.; Shim, W.-G.; Moon, H. Submerged microfiltration membrane coupled with alum coagulation/powdered activated carbon adsorption for complete decolorization of reactive dyes. *Water Res.* **2006**, *40*, 435–444. [CrossRef] [PubMed]
36. Lee, J.-W.; Choi, S.-P.; Thiruvenkatachari, R.; Shim, W.-G.; Moon, H. Evaluation of the performance of adsorption and coagulation processes for the maximum removal of reactive dyes. *Dyes Pigments* **2006**, *69*, 196–203. [CrossRef]
37. Furlan, F.R.; da Silva, L.G.D.; Morgado, A.F.; de Souza, A.A.U.; de Souza, S.M.A.G.U. Removal of reactive dyes from aqueous solutions using combined coagulation/flocculation and adsorption on activated carbon. *Resour. Conserv. Recycl.* **2010**, *54*, 283–290. [CrossRef]
38. Furlan, F.R.; da Silva, L.G.D.; Morgado, A.F.; de Souza, A.A.U.; de Souza, S.M.A.G.U. Application of Coagulation Systems Coupled with Adsorption on Powdered Activated Carbon to Textile Wastewater Treatment. *Chem. Prod. Process Model.* **2009**, *4*, 8. [CrossRef]
39. Lagergren, S. Zur Theorie der sogenannten adsorption geloster stoffe. *K. Sven. Vetensk. Handl.* **1898**, *24*, 1–39.
40. Ho, Y.S.; Mckay, G. Pseudo-second order model for sorption processes. *Process Biochem.* **1999**, *34*, 451–465.
41. Ho, Y.S.; Mckay, G. The kinetics of sorption of divalent metal ions onto sphagnum moss peat. *Water Res.* **2000**, *34*, 735–742. [CrossRef]
42. Dulman, V.; Cucu-Man, S.M. Sorption of some textile dyes by beech wood sawdust. *J. Hazard. Mater.* **2009**, *162*, 1457–1464. [CrossRef] [PubMed]
43. Baccar, R.; Blánquez, P.; Bouzid, J.; Feki, M.; Sarrà, M. Equilibrium, thermodynamic and kinetic studies on adsorption of commercial dye by activated carbon derived from olive-waste cakes. *Chem. Eng. J.* **2010**, *165*, 457–464. [CrossRef]
44. Travlou, N.A.; Kyzas, G.Z.; Lazaridis, N.K.; Deliyanni, E.A. Graphite oxide/chitosan composite for reactive dye removal. *Chem. Eng. J.* **2013**, *217*, 256–265. [CrossRef]
45. Elwakeel, K.Z.; Rekaby, M. Efficient removal of Reactive Black 5 from aqueous media using glycidyl methacrylate resin modified with tetrethelenepentamine. *J. Hazard. Mater.* **2011**, *188*, 10–18. [CrossRef] [PubMed]

MDPI AG
Grosspeteranlage 5
4052 Basel
Switzerland
Tel.: +41 61 683 77 34

Molecules Editorial Office
E-mail: molecules@mdpi.com
www.mdpi.com/journal/molecules